Faculty Development in Nordic Engineering Education

Anette Kolmos, Ole Vinther, Pernille Andersson, Lauri Malmi and Margrete Fuglem, eds.

Faculty Development in Nordic Engineering Education

Anette Kolmos, Ole Vinther, Pernille Andersson, Lauri Malmi and Margrete Fuglem, eds.

ISBN 87-7307-727-5

Copyright 2004 The Authors and Aalborg University Press

Printed by Special-Trykkeriet Viborg a/s

Cover: Eva Sofie Rafn

Published by
Aalborg University Press
Niels Jernes Vej 6B
9220 Aalborg Ø
Phone + 45 96357140
Fax + 45 96350076
E-mail: aauf@forlag.aau.dk
http://www.forlag.aau.dk

This book was made in co-operation with:
Pedagogical Network for Engineering Education in Denmark (IPN)
Niels Bohrs Alle 1, 5230 Odense M
http://www.ipn.dk/index.html - **E-mail:** ipn@iot.dk

UNESCO International Centre for Engineering Education, Centre for Problem Based Learning (UCPBL), Aalborg University, Frederik Bajers Vej 7 A1, DK-9220 Aalborg Ø, Denmark.
http://ucpbl.org – **E-mail:** ucpbl@kom.aau.dk

IPN – series No. 1

Contents

Perspectives

Anette Kolmos, Ole Vinther, Pernille Andersson, Lauri Malmi and Margrete Fuglem:
Perspectives on Nordic Faculty Development..................................5

Erik de Graaff:
A European Perspective on Faculty Development............................13

Gunilla Jönson:
The Breakthrough Project...21

Torgny Roxå and Pernille Hammar Andersson:
The Breakthrough Project –A large-scale project of pedagogical Development..25

Hans Peter Christensen:
Creating a Learning Environment for Engineering Education..............49

National strategies and accreditation

Anette Kolmos, Vidar Gynnild and Torgny Roxå:
The Organisational Aspect of Faculty Development........................67

Pernille Hammar Andersson and Lauri Malmi:
Proposing Nordic Excellent Teaching Practice, NETP......................89

Context

Johanna Naukkarinen and Lauri Malmi:
Faculty Development in Engineering Education in Finland................97

Margrethe Fuglem:
The Educational System within Engineering in Norway –
Development Strategies at Faculty Level...................................111

Anette Kolmos and Ole Vinther:
Faculty Development Strategies at the Danish Engineering
Education..123

Methods

Säde-Pirkko Nissilä:
Reflection for Staff and Faculty Development...............................133

Pernille Hammar Andersson and Torgny Roxå:
The Pedagogical Academy – a Way to Encourage and
Reward Scholarly Teaching..151

Søren Hansen, Säde-Pirkko Nissilä, Claus Spliid and
Anders Ahlberg:
Portfolio – Why, What and How..159

Bertil Larsson and Anders Ahlberg:
Continuous Assessment..169

Chapter 1

Faculty Development in Nordic Engineering Education

Anette Kolmos, Ole Vinther, Pernille Andersson,
Lauri Malmi and Margrete Fuglem

1. Introduction

In November 2002, the Danish Pedagogical Network for Engineering Educations (Ingeniøruddannelsernes Pædagogiske Netværk) initiated the first Nordic conference on faculty development, held at Aalborg University, Denmark. Participants from Denmark, Norway, Sweden and Finland organised the conference. It was the first conference in a number of conferences on improvement of faculty development methods in the engineering educations. In November 2004, the second conference will be held in Odense, Denmark.

These conferences are organised as workshops in which it is possible to share experience and discuss new initiatives. The purpose of these workshops, which have participants from several institutions and countries, is to facilitate further co-operation and to start co-operative work between institutions and countries. This book is a result of this way of organising the conference, where different groups formed at the conference in order to work on analyses and develop the present practice in the Nordic engineering educations.

2. Challenges in developing engineering education

There is a big need to discuss improvements of faculty development methods in the Nordic countries. Today, it is very challenging to teach at engineering educations because teachers have to deal with a lot of different issues at the same time. The classroom is a mirror of society, it might, however, be difficult to change the institutional culture and the content of the educations according to the demands of the society.

The students have more diverse backgrounds than before and engineering students are no longer a well-defined category. The number of international students has increased in particular. This change is, partly, caused by the declining number of potential Nordic students per year and, partly, because internationalisation and globalisation are on the agenda. With more international students the institutions are able to recruit the same number of students as they did before. Most teachers have to teach both national and international students at the same time and more and more university programmes are run in English in order to meet the requirements of internationalisation.

Adult students are another new group in engineering educations. They too have entered the engineering institutions because of the smaller number of young students, but also because it is getting more and more accepted that one of the tasks for higher engineering education is to organise continuing education and work based learning for industry. It is expected that the teachers are able to develop new educational programmes or adapt existing programmes to new target groups. This issue is closely connected to the raise of part-time studying when more and more students cannot attend all regular courses. New methods for organising courses are needed, and there is a constant push to promote distance learning courses, providing course contents in the web and using ICT tools in many different ways.

Interdisciplinary is on the agenda as well – and it is expected that staff are able to develop interdisciplinary programmes. In order to do so, staff have to co-operate cross discipline borders and together quite different types of colleagues develop and improve the curriculum together.

Engineering sciences and technology develop rapidly. Therefore more and more subjects and disciplines move into the engineering educations while only few subjects move out. Thus an overloaded curriculum is a huge problem, which has to be dealt with in order not to demotivate the students. The challenge of teachers is to identify and stress the important long lasting methods and techniques while not overlooking currently used technologies.

There are many pedagogical problems in engineering educations, but the most important issue is the question of what to do to counter these problems.

Staff and faculty development is one of the answers, but what kind of national and institutional strategies are used? Which methods are used and what attitude to pedagogical development do the managers have?

The contributions to this book are gathered under three headlines:

- Perspectives, which contains more general perspectives on faculty development on institutional, national and Nordic level.
- Context, which contains descriptions of the engineering educations and the specific faculty development methods used.
- Methods, which contains discussions about specific methods used.

3. Perspectives

At many universities all over Europe, the teachers have to develop their pedagogical competence by their own experiences based on trial and error. Trial and error can be a very efficient way of learning at an individual level, but it is a slow learning process. If changes are of more general character and take place at a system level, organising relevant staff development activities is a more efficient way for organisations to move forward.

This has been done in Scandinavia. Norway and Sweden were the first to start training of teachers. Denmark and Finland started training during the 90'ies.

Arguments for staff development are, however, still needed. Erik de Graff turns to the European perspective, among other reasons because of experiences with faculty development at Delft University, the Netherlands. He gave one of the keynotes at the conference and argues in his article, *A European Perspective on Faculty Development*, that the large increase in the number of students has effected and changed the conditions for higher education resulting in mass higher education and that the attempts from the EU to unify educational systems have a tremendous impact on the methods of teaching in higher education. As a consequence, there is an increasing need for staff development.

Gunilla Jönson, rector of Lund Institute of Technology, Lund University, gave the other keynote at the conference. She underscored the importance of management in the development of pedagogical strategies. At Lund Institute of Technology, a large institutional project, *The Breakthrough Project*, has been carried out. This project focused on 5 elements in staff development: Pedagogical courses, Consultative support, Rewards for professionalism in teaching, Evaluation of undergraduate teaching and Research into educational development as well as teaching and learning.

In their article, *The Breakthrough Project – A large-scale project of pedagogical development*, Torgny Roxå and Pernille Hammar Andersson give very elaborate reasons for why the different elements were included and explain the specific methods used in the project. It demonstrates how important it is to use more pedagogical development methods at the same time if a general institutional development, including a change of the culture, is the aim of a project.

In his article, *Creating a Learning Environment for Engineering Education*, Hans Peter Christensen, underscores that it is important to think holistically, as is specifically done in The Breakthrough Project. The aim of the article is to develop a model for learning environment, which contains the student's, the teacher's and the institution's perspective. Faculty development must be improved with a view to considering all three levels.

4. National strategies and accreditation

The staff development strategies in the Nordic countries have a lot in common: In Norway, Sweden and Denmark staff development is compulsory. The requirements are worded in different ways, but the key point is that there is a formal requirement for university teachers to qualify in this area.

It is a general characteristic in the three countries involved, that the strategies for faculty development are **national decentralized strategies**.

In their article, *The Organisational Aspect of Faculty Development*, Anette Kolmos, Vidar Gynnild and Torgny Roxå compare the national strategies in Denmark, Norway and Sweden. In the article, it is particularly pointed out that the faculty development centres ought to be much more conscious in their organisational strategies and focus on all levels of the organisation. Without support from the management it is difficult to get through with pedagogical changes. The common problem in all the Nordic countries is that faculty development not automatically is research-based; this is pointed out as well. One result of this is a lack of evidence of the effect of the pedagogical training.

There is no common national certification level which for example has been developed in Baden Württemberg and Bavaria, Germany. The certification takes place at the individual institutions. There are a lot of advantages by having institutional programmes and letting the institutions develop their own curriculum for staff development. Institutional ownership, motivation, influence and development of relevant institutional profiles are just some of the advantages. However, the disadvantages, such as the dependence on the management's strategies which may change, are unfortunately also important. Another disadvantage is the large differences in the certification processes of the various institutions with regard to both content and level. To some extent programmes may be well known within a country. However, it may cause international problems in the long run, because there will be too many and too specific certification processes – and there is a need for accreditation.

Erik de Graaff argues that it is time for more seriously to discuss certification of teachers in HE and accreditation of HE teacher training programmes. Pernille Hammar Andersson and Lauri Malmi argue for starting the development of a Nordic Excellent Teaching Practice, *NETP*, with the objective to promote and evaluate teaching skills of engineering educators in Nordic countries. NETP would be based on preparing teaching portfolios, which would be evaluated using common criteria and similar processes in Nordic technical universities and institutions of engineering education.

Whether it should be a common NETP or accreditation of existing programmes are open to discussion, but the important thing is that we are at a stage where something has to be done to ensure and support the development of and quality in higher engineering education.

5. Context

This part contains three descriptions of the national engineering educations and faculty development strategies. Two articles, Johanna Naukkarinen and Lauri Malmi's *Faculty Development in Engineering Education in Finland* and Margrete Fuglem's *The Educational System within Engineering in Norway*, take a historical perspective of engineering educations in their respective countries as a starting point.

In the article by Anette Kolmos and Ole Vinther, *Faculty Development Strategies at the Danish Engineering Education*, the structure of engineering educations in Denmark, the staff development centres and the formal requirements to staff development are described.

The situation in Sweden is indirectly illustrated in two articles in the previous part and the next part of the book.

6. Methods

In the last part of the book, the focus is on specific faculty development methods. The methods are varied and the different methods have different advantages and disadvantages.

In her article, *Reflection for Staff and Faculty Development*, Säde-Pirkko Nissilä focuses on a more general methodical aspect, namely that it is important to make the teachers' and the teacher trainees' tacit knowledge explicit, and to facilitate their ability to learn from experience and theory by systematic reflection.

In Pernille Hammar Andersson and Torgny Roxå's *The Pedagogical Academy – a Way to Encourage and Reward Scholarly Teaching*, there is a concrete example of the use of reflection and portfolios. The main objective of the pedagogical academy is to encourage pedagogical development at Lund Institute of Technology and to reward good, ambitious and quality-conscious teachers. The article describes very concretely how the methods are used.

Søren Hansen, Säde-Pirkko Nissilä, Claus Spliid and Anders Ahlberg discuss in the article: *Portfolio – Why, what and how* the role and function of teaching portfolios in engineering education. The portfolio method is used more and more often among teachers in Nordic engineering education. But why has this method become so popular? What is a portfolio and how can they be implemented in the system?

In the final article, *Continuous Assessment*, Bertil Larsson and Anders Ahlberg emphasise that formative assessment is a tool for quality development. Formative course assessment should be organised as a productive dialogue between teachers and their classes. From the schools' perspective, it is therefore imperative to monitor good opportunities for the teachers to continually develop their understanding of student learning and how teachers manage it. The final aim of faculty development in the area of pedagogic is to improve the students' learning and to educate future engineers with excellent skills.

6. Enjoy the book

There are great challenges to work within this area of pedagogical development in engineering education both at a local level in the institutions and the universities, and at an international level. In the latter case, it is a matter of developing systems of good co-operation, to take advantage of different experiences and develop a quality assurance system. We hope this book, showing the present state of Nordic and national faculty and pedagogical development in engineering education, can spread some new insights and inspires to further development in this area.

Chapter 2

A European perspective on Faculty Development

Erik de Graaff

1. Abstract

According to European traditions in higher education a professor is primarily a scientist and a researcher. Teaching used to be closely related to the field of research and in most cases a rather small-scale operation. The large increase in student numbers has resulted in mass higher education. Added to the diversification and specialisation in science this imposes changes on the system of higher education and consequently on the methods of teaching. As a result, teaching in higher education is developing into a profession in its own right. At the same time, influences from the political front of the expanding European Union incite a general move towards converging of the different national systems of higher education. The paper highlights some major issues of concern for faculty developers and educational innovators like the identification of relevant competencies for teaching in higher education, certification of university teachers and accreditation of university teacher training programmes.

2. Introduction

In most European countries someone who teaches at a university is addressed as professor. This suggests a relationship with the concept "profession". In particular in engineering education this connection seems most apt. Professors in engineering are a relatively new phenomenon. Until the 19th century a scientist usually covered a broad field of interest, like natural sciences, law, philosophy and theology. In his field the professor conducted his investigations and published on his findings. In the course of his investigations a professor in natural sciences sometimes conducted experiments. In the course of such experiments the professor had to perform the engineering task of designing and constructing his own instruments. This way Galileo build his own telescopes that enabled him to see the stars in the Milky Way and the German physicist Otto Von Guericke who invented the air pump in the course of his famous experiments with vacuum inside the two hemispheres in Magdeburg, to mention just two examples. An interesting detail is that as a reward for his achievements Galileo was discharged from his teaching duties so that he could devote himself completely to his investigations.

So scientists were engineers when it was needed and there was no need for a professor in engineering. Gradually, this has changed as part of the ongoing specialisation in science.

Even if professors were appointed in engineering fields before, in Holland for instance, it has lasted until after the second half of the last century for engineering schools to gain the status of a University of Technology with the right to appoint a staff of full professors. Traditionally, a professor in engineering needs to be an expert in his field of practice. As such he can set an example of professional behaviour. Most professors in engineering consider this to be the quintessence of teaching. Therefore, it is easy to understand that most professors feel there is little need for didactic training (just as is the case with most professors in the general universities).

In this paper it will be argued that over the past decades the task of teaching at a university has changed considerably, and that presently didactic training is definitely a prerequisite for anyone teaching in higher education. Next, different strategies for implementing a faculty development programme will be discussed. The case of the Delft Faculty development programme will be presented as an example. Finally the future of faculty development in engineering education will be discussed.

3. Present needs for faculty development

Universities have to support students to acquire new knowledge and higher order cognitive skills to enable them to adapt to new contexts and pursue learning, whatever the conditions (Prosser & Trigwell, 1999). Since World War II the system of higher education in Europe has gone through a series of significant changes. For one thing, a growing part of an increasing population participates in one or another form of higher education. No longer is it only the sons of the elite that go to university. Their sisters have joined them, and also a host of people from less privileged classes. As a consequence of the increasing number of students, the old individualised teaching model has become obsolete. In most European universities a freshman class nowadays consists of several hundreds of students. The professors, teaching in large lecture halls, need amplification simply in order to be heard by all students. Evidently, didactic skills play a different and more important role in such a large class than it did in the old model of private tutoring. When we adhere to the assumption that good learning depends on good teaching (Biggs, 1999), it should pay to invest in improving teaching competencies.

A second factor, which increases the need for didactic training, is particularly important in engineering education. It concerns the ongoing diversification and specialisation in science and the increasing speed of technological innovations. As a result of the rapid technological development, teachers in higher engineering will find themselves in a situation where they are increasingly unable to deal with questions from students directly, because their own experience has become outdated. Moreover, the students will have easy access to the most recent information online, and they will confront teachers with this knowledge. Consequently, the traditional teaching method of transferring knowledge from the experienced person to a layperson will fail more and more often (if it ever was very successful at all).

In its turn, this leads to the third factor. Teachers in higher education will have to learn to deal with new educational methods, which put a higher emphasis on the students' abilities to direct their own learning process, like problem based learning (PBL), experiential learning and project organised learning (De Graaff & Kolmos, 2003). This means that they have to acquire competencies related to new teacher roles, like facilitator skills, advisory skills, etc. At the same time the teachers will have to acquaint themselves with new teaching technologies ranging from Internet software to simple presentation tools. Hence teachers in higher education will have to become life-long-learners.

A fourth factor, relevant in relation with the training of teachers in higher education, is the internationalisation of tertiary education. As a result of the increased mobility of the people, universities are now competing across the national borders. In Europe the Bologna declaration has reinforced this process. The endeavour to unify the European countries' systems of higher education promotes international student exchange. In many institutes the master programmes are now being taught in English, the lingua franca of the modern world. For the professors this poses yet another challenge to their didactic skills. It has been demonstrated that for teachers who are not native speakers of English the didactic skills become more important (Klaassen et al., 2003). No matter how much effort you spend in improving you language skills, your command will always be less than perfect, and this will hamper your ability to interact with the students.

Summing up we may conclude that didactic skills have become more and more important in higher education all over the world. Teaching at a university is developing into a profession on its own accord. Both for the professors as well as for the institutes it has become necessary to clarify the demands for qualification as a teacher in higher education. So, the implementation of a faculty development programme has grown to be an urgent issue.

4. Faculty development strategies

The implementation of a faculty development programme presupposes an institutional culture where teaching activities are considered to play an important part. Academic leaders have a prominent role in this sense. Several authors stress the importance of institutional recognition of the quality and value of teaching in higher education by academic leaders at all levels (Wright, 1995; Knight & Trowler, 2001). All institutional policies and practices regarding teaching have to be fully supported by academic leaders from the lowest to the highest levels. Demonstrating institutional commitment can take many forms from providing financial support to the organisation of special events, initiating pilot programs, opening workshops and handing out certification at the end of programs, etc.

Institutionalising a Faculty development programme is just one of the prospective actions. The summation of possibilities indicates that in order to be successful a strategy to implement Faculty a development programme, it should be embedded in a "culture of teaching", and it should be part of a strategy. In a meta-analysis of various faculty development strategies (Rege Colet, 2002) points at three basic Faculty development strategies:

- A programme of courses and workshops offering teachers opportunities to improve themselves.
- A teacher training programme leading to a formal certification.
- Continued education of teachers as life long learners in a learning community

The three strategies build up to an increasing complexity and involvement. Also they more or less represent a developmental order. The first approach has been applied in a good many universities over the past thirty years. An obvious advantage of the approach is that it extends an open invitation to university staff to improve their teaching skills. The other side of this medal is that often the right persons are not reached (see the account on the faculty development programme in Delft in par. 4.). A little more pressure form the university administration in combination with some rewards attached to an increase in teaching skills might make all the difference.

The second type of strategy aims to achieve this by awarding a certificate for demonstrated teaching competencies. The certificate can be used as a condition for promotion or other rewards. A qualification programme should include restructuring of the teachers' knowledge, practice and the production of validated knowledge on teaching and learning. (Tillema & Imants, 1995). As (Rege Colet, 2002) points out, there is not a big fundamental difference between the two approaches. A common basic disadvantage is that the procedures still suggest that teaching is a limited set of skills that you can learn once and than apply for the rest of your life.

The objective of the third strategy is a more fundamental change. The goal is set at creating learning communities of teachers. Through Intervision and supervision techniques, teachers can learn form each other. Participation in these activities may change during one's career, but it will never be finished. The third strategy appears to be ideal. However, it has the characteristic disadvantage of an idealist system. In order for it to work this system needs voluntary participation. Therefore the risk exists that, just like with strategy 1, only those persons will be involved who were already teaching- minded to begin with.

5. The Delft University teacher Qualification Programme

The following section will relate to the experiences with the development of a faculty development strategy at Delft University of Technology. In the seventies of the last century, interest in educational development and teacher training started to grow everywhere in the Netherlands. In Delft, like at all other Dutch Universities, a centre for educational development and didactics was founded. During the following years this centre (DIDO) builds a programme with courses for university teachers. The courses were open for enrolment for all university staff without a fee.

Those who participated in the courses were usually quite satisfied with the programme. However, the number of professors participating in the programme remained relatively low. In particular some professors that were thought to need the courses most (judging by the comments from the students in their annual feedback reports) never did show up at all. So the programme was not succeeding in reaching the right population and also the effect of the isolated courses was estimated to be of limited value. Therefore the development of a new programme for teacher qualification was initiated in the strategic policy planning for the years 2000-2003 (Klaassen et al., 2003).

The new TU Delft Teacher Qualification Tract for apprentice university teachers is designed with focus on the development of didactic competencies. Central to the programme is a teaching competence portfolio. At the start of the tract each participant is interviewed to establish his/her entry level, resulting in a Personal Learning and Development Plan. In this individual study plan the participant can choose from a modular programme offered by the Didactic Group at TU Delft. The modular courses allow the participants to follow only those courses that are needed to acquire the missing competencies. If they like, the participants can choose other means to achieve the necessary competencies. In any case, they all have to work on completing their teaching portfolio, which constitutes the basis for the final assessment and the grounds for certification as a qualified university teacher. Starting with an entry level of a university degree in engineering and little or no teaching experience the total workload, including teaching time is estimated at 250 hours. The maximum time-span to complete the Qualification Tract is estimated at two years.

The modular programme of the TU Delft Teacher Qualification Tract consists of six thematic modules, each consisting of a workshop (2-3 half day sessions), a master class (2 half day sessions), an Intervision session, practice assignments and individual work.
The modules cover the following themes:

- Activating students in a lecture
- Student centred learning (PBL)
- ICT in higher education
- Assessment of learning results
- Teaching in English
- Educational research

During 2002 and 2003 about 40 university teachers entered the Didactic Qualification Tract at TU Delft. So far the experiences with the Tract are positive. At the same time the limitations of the initial "Basic Certificate" are being recognised. Several participants have already indicated that they want to continue after completing the Tract. Also, there are requests from experienced teachers who want to be involved in some sort of teacher training. Within the framework of the modular programme, part of these requests can be met. However, the next step will be the development of a specific programme and added certificates for senior teachers.

6. Discussion

Over the past years, attention on training and qualification of university teachers has increased evidently. Still there is no agreement on a European level as to which competencies are needed and how they can be assessed uniformly. In order to be allowed to teach at kindergarten level a candidate must spend a serious amount of time on didactic training. To teach in university it still suffices to be an expert in a specific field. The three types of strategies that were discussed each have their own weaknesses. Strategies 1 and 3 are both vulnerable to a lack of motivation which makes it difficult to attract the right people. Strategy 2 aims at providing an external source of motivation. The risk of this approach is that people lose their interest as soon as the reward is collected. In order to solve this puzzle a combination between strategies 2 and 3 might prove to be the most effective. Qualification and certification are necessary to raise the status of teaching; however, the learning process should not stop a basic certificate.

In the course of the Bologna process we need to define a European standard for teacher qualifications in higher education. Such a standard should be based on clearly defined teaching competencies. With the criteria attached to this standard it should be possible to differentiate between the entry level, a basic teaching qualification and surplus qualifications. In order for these qualifications to be recognised in the different European countries accreditation procedures need to be set up. Hence, there is still a lot of work to be done before we can all join the international "community of practice" of teachers in higher education.

References

Biggs, J. (1999) *Teaching for Quality Learning at University,* Society for Research in Higher Education and Open University Press, Buckingham.

Graaff, E. de & Kolmos, A. (2003) Characteristics of problem-based learning. *International Journal of Engineering Education.* Vol.19, No.5, pp 657-662.

Knight, P. T. & Trowler, P. R. (2001) *Departmental Leadership in Higher Education*, Society for Research in Higher Education and Open University Press, Buckingham.

Klaassen, R. G. & Graaff, E. de (2001) Facing innovation: preparing lecturers for English-medium instruction in a non-native context, *European Journal of Engineering Education* Vol. 26, No 3, pp 281-289.

Klaassen, R. G., Andernach T. & Graaff, E. de (2003) A Qualification Programme for University Teachers in Engineering, In: Alfrede Soeiro & Carlos Oliveira. *Global Engineer: Education and Training for Mobility.* Proceedings of the 31th annual SEFI conference, 7-10 September, Porto, pp 126-130.

Prosser, M. & Trigwell, K. (1999) *Understanding Learning and Teaching. The Experience in Higher Education.* Society for Research in Higher Education and Open University Press, Buckingham.

Rege Colet, N. (2002) *Faculty Development Strategies.* Issue paper presented at the Stratan Etan expert group on the relationship between Research and Higher education, Brussels.

Tillema, H. H. & Imants, J. G. M. (1995) Training for the Professional Development of Teachers, In: Thomas Guskey & Micheal Huberman, *Professional development in Education*, Teachers College Press, New York.

Wright, W.A. et al. (1995) *Teaching Improvement Practices: Successful Strategies for Higher Education.* Anker Publishing, Bolton, MA.

Chapter 3

The Breakthrough Project

Gunilla Jönsson

1. Background

Institute of Technology at Lund University operates within a competitive environment. In a society that constantly asks for more from its higher education sector. New student groups are admitted to higher education, their background, values and conceptions of knowledge are no longer the same as they used to be. Through the government, the taxpayers ask for value for money and real contributions to the growth of welfare. They expect higher education to contribute significantly to this. In addition to this there are demands from the sector itself. There is a need to formulate and defend the values that governs knowledge construction within a scientific community.

Research has always been highly competitive and has since long constructed its own rules and arenas. Lund Institute of Technology claims a high profile in this area and reaches even further. But, signs are showing that an institution in higher education can no longer trust excellent research only as the means to defend its position, neither in society nor in the scientific community. The objectives have to be aiming even further. An institution has to be excellent not only in research, but also in both teaching and its administrative functions. The focus of this text is how to endorse excellence in teaching.

On the Swedish arena there is a shortage of students prepared for further studies in technology and engineering. There are several reasons for this. One is the expansion of the sector. More seats are made available than the secondary school manages to provide students for. From this follows a need for institutions to prove that they are able to support student learning. If they can, and this is made known to future students, these students are more likely to choose that institution before others. So, how can this be done? How can an institution claim a serious attempt to raise the standards of teaching?

At Lund Institute of Technology several initiatives have been taken in order to be able to claim seriousness in its strive to improve teaching and student learning. It would be hard to prove that its ability to support student learning exceeds other institutions in Sweden. It would be hard for anyone since measurement of such matters is too difficult. Instead the objective is to build a strategy for improvement, which qualities are made clear to everybody.

The Breakthrough project was formulated as such an initiative. It rests on five tracks for development. Pedagogical courses for teachers, consultative support for staff where they actually do their work, a reward system for teachers and departments showing professionalism in teaching, a system for evaluation providing feedback to the system and to individuals, and a nurturing of research on educational development and learning in order to describe what is going on more systematically and also in order to be able to contribute to the scientific society as a whole. The overall objective is to make Lund Institute of Technology known as an institution taking its teaching seriously and to improve the capacity to support student learning. Reaching this, hopefully more and more qualified students will look for Lund as a promising alternative and this would benefit everybody.

Pedagogical courses Teachers in higher education and maybe in engineering and technology especially, usually have no formal education to become teachers. If they are about to improve teaching the institution has to provide new and existing teachers support in terms of pedagogical education. They aim to raise their ability to understand what is going on in the teaching and learning situations they observe. By raising their understanding they may make better decisions and carry out better teaching, all in order to support the students in their learning task. All new teachers will consequently have ten weeks of pedagogical education and all present teachers are encouraged to participate in the courses offered.

Consultative support When the demands for quality in teaching are raised the institution provides pedagogical support for all levels within the organisation, from individual teachers to the Dean's office. The support is provided from recourse persons already employed within the organisation and specially trained for their task. They are supervised and supported by experienced pedagogical consultants.

Rewards for professionalism in teaching A system where individual teachers are given a title and higher salary for professionalism in teaching is implemented. In order to reward not only individuals but also to encourage structural development the departments where the teachers are active also get rewarded with increased funding. The system is based on peer assessment of teaching portfolios and interviews.

Evaluation of undergraduate teaching A faculty wide system for summative evaluation of undergraduate teaching is implemented. It uses the Course experience Questionnaire (Ramsden, 1996), but is not to be considered marking of teaching. Instead the core of the system is the comments made by teachers, students and program directors and enclosed to the reports. The aim is to inform the critical conversation within the institution dealing with teaching and learning. The summative evaluation is supplemented by a formative evaluation done during courses as an intensified dialogue between teachers and students (Angelo & Cross, 1993).

Research into educational development and teaching and learning In order to learn more and to gain a position in the academic community also concerning educational development, the institution nurture smaller research project investigating effects and processes of its various initiatives. This is important in order to improve even further and to ground this improvement in scientific evidence. But, it is also important in order to win and maintain the institution a position as leading the development of teaching and learning within engineering and technology education.

2. Experiences gained

No doubt the institution is strengthening its ambitions in an area slightly alien to its scientific specialities. Despite this tremendous progress has been made. A majority of the teachers have engaged and engage in a continuously improved discussion about teaching and learning. Curriculum is changing as a result of this. Lund Institute of Technology has begun to earn a position on the national arena and has earned a great deal of interest and respect on the international. Yet there is a great deal to be done and the institution is determined to continue.

A large university faculty has been described as a landscape divided into many territories and inhabited by many tribes (Becher & Trowler, 2001). This is also true for this institution. As the conversation intensifies the different points of views are elaborated and the individuals defending them invest more prestige into them. This makes it absolutely clear that an initiative like this rests on respect for different perspectives. The informed conversation is the only compass to follow while developing practice in an academic environment. It builds on the participants' ability to formulate arguments grounded in research and supported by evidence. This is also the experience gained during the process described above.

Evidence in the area of teaching and learning might appear illusive, especially to people used to deal with hard facts and mathematical models. Even so, an institution like Lund Institute of Technology must learn to master it. The resources of the last century will not return, so the only way to improve is to utilise resources in the best possible way. Student learning can be described as the outcome of all teaching. If we are to improve our results we have to learn to be more effective in the way we support the students during their studies. The result will earn us the reputation we can build on in the competitive environment where we operate.

References

Angelo, T. & Cross, P. (1993) *Classroom Assessment Techniques*, Jossey-Bass.

Becher, T. & Trowler, R. (2001) *Academic Tribes and Territories*, The Society for Research into Higher Education and Open University Press.

Ramsden, P. (1996) *Learning to Teach in Higher Education*, Routledge.

Chapter 4

The Breakthrough Project – A large-scale project of Pedagogical Development

Torgny Roxå and Pernille Hammar Andersson

1. Abstract

Today, it is necessary for higher education institutions to change many traditional ways of handling processes in their organisations, as well as, how teachers handle and look upon students' learning and their own teaching. In Sweden, this development is driven by changes in the society, which has led to new groups of students entering the universities with a much more diverse educational, social and ethnical backgrounds than just ten or fifteen years ago. In Sweden, it is an aim to recruit these new groups of students to higher education.

To solve this situation in a professional and adequate way, it is critical that the teachers increase their competence in teaching and learning and their ability to handle and understand the students' learning situation. The Universities have to face these new challenges, develop new strategies to meet them, and to take advantage of the new possibilities they are bringing.

The pedagogical research has increased our knowledge of teaching and student learning in higher education. This factor also has an effect on teaching and learning. Important are also the new demands on the students' skills as employees when they enter the labour market after their education. The impact this has on higher education is that these skills have to be trained during the students' time at university. The body of knowledge is also growing and, as a consequence, the complexities of the subjects the students are supposed to learn are increasing. At the same time, the amount of funding is decreasing in Sweden at present.

All these factors are well known to everyone working in higher education; still, it raises the demands to find new solutions when the traditional ways not are working effectively anymore. It is necessary to start developing new professional ways, in higher education, to handle teaching of students and to handle the students' learning (Bowden & Marton, 1999).

In order to deal with those challenges, the Breakthrough project started at Lund Institute of Technology, the faculty of technology at Lund University in 2000. In this paper we will describe this large-scale project and its assumptions.

2. The Breakthrough Project

The basic idea of this large-scale pedagogical development project at LTH (Lund Institute of Technology) is that all development activities must be based on an understanding of people's interactions. Therefore, all development activities must take into accounts the possibilities and constraints emerging from a complex social system. Learning is closely connected to the identity of the individuals involved (Wenger, 1999). They may very well have individual and unique knowledge, but all actions taken within a social context will be regulated by the responses given by other members of the community. Those social processes are important starting points in every large-scale project.

The project at LTH *The Breakthrough Project* rests on a conviction that the teachers already strive for quality in their teaching and also try to improve their teaching continually. They are committed to being teachers and most of them are proud of teaching the students. Despite of this, the system has been fairly stable when it comes to taking care of the students' learning. This paper describes some of the initiatives taken in order to start a process of making these efforts to development more stable, conscious and striving in the same direction. It also, as a start, considers some of the factors that are found to be important while searching for constraints working against a continuous development.

It is important to mention that this is a description of the features observed during the work with this project and also a description of the underlying ideas what can be of importance when developing academic organisations. The description is not to be seen as a scientific work or statement. There have been no scientific investigations to support the statements in this article. It can be seen as a sharing of the experiences made.

The Breakthrough project has to deal with an institute of approximately 6000 students, 500 senior teachers and about 500 postgraduate students who almost all do some kind of teaching. The students are spread at 17 graduate programs, with duration of four and a half years, and 11 engineer educational programs, with a duration of three years. At LTH, there are also a School of architecture, educations in industrial design and risk management and a Fire Safety Engineering program.

The Breakthrough project was a three year project with an annual budget of 300 000 €. It has been running since 1st January 2001. Since the beginning of 2004, it has been a permanent competence development activity at LTH.

2.1. Goals and objectives

The Breakthrough Project is supposed to make LTH a faculty of technology that systematically and consciously develops its teaching and learning capacity (knowledge) and strategies. Another important goal is to make LTH well known for doing so. This is supposed to happen by increasing the competence of teaching and learning at the whole faculty but mainly for the teachers. In this, there is a striving for the teachers to develop learning focused education. It is also important to try to create a culture with a higher extent of collaboration concerning students learning. This corresponds to Boyer's theory of Scholarship (Boyer, 1990).

2.2. The background for the project

There are many reasons for developing teaching and learning situations at a faculty of technology. The reasons for this project are the changes universities and higher education face and that already are affecting them. Knowledge has a new role in society, new groups of students are entering the universities with more diverse backgrounds than before, more people in different ages are supposed to go through higher education, and, at the same time, the funding from the government is decreasing. All new challenges are supposed to be met with fewer resources than before.

In Sweden, it is an additional concern that not enough young people are interested in higher education. Particularly, there are too few interested in science and technology. At LTH, this is another reason for trying to increase the interest in issues of teaching and learning to the same level as the interest of issues in research and to try to make this well know.

Experiences from previous activities at LTH have influenced the work in The Breakthrough Project. Pedagogical courses teachers have been important forums where teachers have met to learn about teaching and learning. They have also been given opportunity to discuss what needs to be done to make the student learning even better at LTH. Previous experiences from smaller projects to develop the teaching and learning culture have also shown the direction for the project.

2.3. Project as development strategy

Because it was impossible to foresee the rate of success, it was decided to start with a limited project. Furthermore, it is an advantage to start a project, where the activities not are part of the normal activities in the organisation, because it is not regarded as a threat to the activities everybody is familiar with. It is also easier to highlight the activities in the project.

There are good reasons for forming a project organisation that not are part of the normal activities. However, it could also be claimed that it makes it harder to incorporate new ways of working and thinking. The Breakthrough Project is a project working with the development of processes in the ordinary activities. The goal with this way of working is to train the members of the organisation to take responsibility for their own development.

A project is also a kind of learning situation. Development is often learning new ways of seeing and interpreting one's actions and the surrounding world. One model of learning from experience is Kolbs circle of learning (Kolb, 1984). It can be used to describe the philosophy of the working methods in The Breakthrough Project.

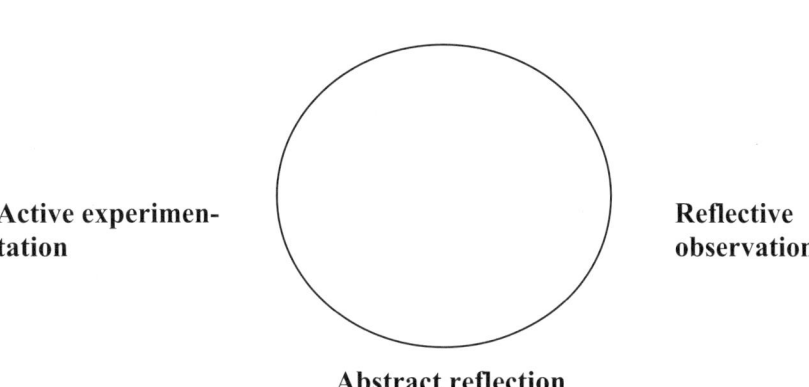

During the phase of Concrete experience, people in the organisation experience that something new is happening through actions and reasoning from the leading persons at LTH and through different activities that The Breakthrough Project is arranging. This makes people think, reflect, and pose questions about their experiences and what they mean. As soon as someone wants to go deeper into their reflections, and maybe develop a new way of teaching, a consultant from The Breakthrough Project offers support in the process. This means that the Breakthrough Project supports development through the phases of Abstract reflection, Active experimentation and, hopefully, it comes to a real change in action that leads to a new Concrete experience.

3. Main features of the large-scale project at LTH

As a start, it is important that all organisations in higher education find out what will make students' learning and study situations better and what can create satisfaction for teachers in their teaching situations. It is about creating a good working environment for both teachers and students.

It is also a mater of finding the strengths and weaknesses of the organisation and trying to formulate goals and reasons for starting a development process. It is also important to make sure that those arguments are well known in the organisation and to have a continuous dialog about the goal and the process between a lot of different groups and levels in the organisation.

3.1. Scholarship of teaching

One of the goals for LTH is to invent a culture of Scholarship of teaching. In his book *Scholarship Reconsidered. Priorities of the professoriate*, (Boyer, 1990), discusses the role of an academic. The issue is to discuss what constitutes a professional. The thesis put forward is that four tasks are essential for an academic to engage in. Three of these are the scholarship of investigation, integration and implementation. The fourth is the scholarship of teaching. Since Ernest Boyer published his book in 1990, a lot of authors have followed in his footsteps and have written about scholarship of teaching. (Healey, 2000; Kreber, 2000; Trigwell, Martin et al., 2000; Kreber, 2002).

Within the Breakthrough project, scholarship of teaching is understood as university teachers' way of engaging in teaching like it is natural for them to do in research. This means that they should investigate their own teaching and its results, discuss their findings and experiences with colleagues and by doing this strive for an ongoing development of their own teaching competence. Teaching is partly a practical activity; therefore the developed knowledge also has to be implemented in practice.

At LTH, the overall goal is to stress that teaching and student learning are as important as research to make LTH a successful institute of technology.

3.2. From a teaching to a learning paradigm

The other main issue for LTH is to shift from a teaching to a learning paradigm (Barr & Tagg, 1995; Prosser & Trigwell, 1999). A paradigm must be understood as something which functions as an imperative on the people working within it. It governs how to do things, how to interpret phenomena and how to discuss. It is the frame of thinking (Kuhn, 1992).

The teaching paradigm focuses on the teachers' actions and the content of the subjects, instead of focusing on the students' learning process and what actually makes good learning. (Barr & Tagg, 1995) argue that the learning paradigm is more suited to make higher education successful under the new demands and the whole, present situation in higher education.

At LTH, like in many universities, the traditional teaching process starts all new parts of the curriculum with an introductory lecture. The students then practise on things presented in the lecture. Many programs also use laboratory work as an additional opportunity for the students to practise. A course (about 5-7 weeks long) ends with a 5 hour written exam. A rough analysis on the underlying idea about teaching is that first the students hear about things, and then they practise on the things they have heard, and in the end their ability to do the tings taught is assessed. This philosophy of learning may very well be described as systematic a master apprentice one. The master shows and all the apprentices try to imitate. There is not so much concern about how the students actually learn and how they handle and understand the task they are given. In other words, there is not much focus on student learning, directly.

It must be stressed that this is a general description. There are plenty of exceptions. However, these exceptions still do not construct the overall picture of teaching at LTH. The desirable situation is that new, learning focused ways of teaching become the average.

It is important to remember that we do not know which knowledge there will be a demand for in the future. That is why we in higher education must be aware of our responsibility to train the students as learners (Bowden & Marton, 1999).

3.3. Pedagogical competences

Of course it is very important to develop the pedagogical competence in the whole organisation but it is most important to develop the teacher's pedagogical competence. The teachers constantly need to develop their competence when it comes to understanding and analysing what goes on in teaching and learning situations. Increasing this competence will make the teachers more competent to take effective decisions and actions about how to teach in a way that support the students' learning better.

Competence is here understood as an ability to analyse teaching and learning situations, design teaching accordingly, carry out the teaching and evaluate the result effectively. This should be done, preferably, as a conscious process in which the experience is shared and discussed with colleagues.

3.4. Higher extent of co-operation

One strong belief governing this project is that a higher extent of co-operation among the teachers is needed. The different actors involved in the education of the students do not always work effectively together. Teachers in the same education program do not always know each other enough and therefore one teacher sometimes carry out teaching that is counterproductive to another teacher's. The result is frustrated and bewildered students, which often lead to extra workloads for the teachers.

Often teachers work isolated from each other and merely regard each other as colleagues. This sense of loneliness and the feeling of sometimes to be in an exposed situation do not create an optimal situation for learning and developing. In the end, this affects the student's learning outcome. The teachers have no one to discuss educational issues and problems with. They often have no one to share their experience with. New teachers are on their own and have to try and learn how to teach the hard way. It is not difficult to understand that many teachers are reluctant to try new ways of teaching if there is no sense of support. Another problem with this culture is that it is not always possible to learn from experiences gained.

Much will be gained if the cooperation among teachers is increased and if they get adequate languages to express their thoughts and experience. Together, these two critical areas make an important ground for creating a culture in which pedagogical issues are handled in a more professional way. Development of the pedagogical competence gives the organisation an appropriate language and common knowledge to discuss teaching and learning. A higher extent of cooperation provides more opportunities to have a dialogue, and experiences of having colleagues and a feeling of confidence. This is one of the most fundamental conditions for learning- and developing processes.

Many philosophers, educationalists and psychologists have since Socrates and Plato claimed that the true dialogue is important for people's development. Through the dialogue between people regarding themselves as equals the individual is becoming aware of oneself as a person among other persons. The dialogue is necessary to make us conscious and self-aware (Egidius, 2002). Interaction between people could often be prevented by a social process that is more about power and hierarchy than listening. This does not lead to development because the most necessary conditions are missing. That is why it is so important to try to create good conditions for teachers and other actors in the organisation, so they will be able to participate in constructive dialogues.

3.5. Goal-oriented management

What are the best ways to work with change processes in an organisation as a university where the employees are fairly free to make their own decisions and control their own work? In a way, this is an important condition in universities. It is important that researchers can find their own ways in their work with subjects to find new creative solutions and new ways of handling problems and scientific questions. However, the new role universities have in society, to be a place for everyone and the knowledge produced are supposed to be used and spread in a more active way, increases the need for some kind of management to make sure this new role is being taken care of in an appropriate way. The solution must be to find ways to work with development in a particular direction and still keep the sense of freedom. Actually, this is important for many organisations in today's modern society.

It is important to find the most critical and central points to develop that also have possibilities to be integrated in normal activities in the organisation. Organisations are to be looked upon as systems in the sense that they are not just the sum of different parts but also a structure. If one part of the structure is changed, all other parts are affected in some way. The point is that this structure, that is an organisation, has a tendency to remain stable and in balance just like the human personality. The consequence is often that changes are rejected because it is natural to remain stable and changes can be regarded as threats of the unknown. Smaller changes are not very helpful in this process even if they work in the same direction. It is critical to begin with developing areas in the organisation that are so central or important that they can influence other parts as well. If these areas integrate changes in a natural way, they thereby contribute to an overall cultural change (Fromm, 2000).

In the Breakthrough Project, the main idea is that the board of the faculty and management at LTH are deciding the overall goals of the organisation. They are not presenting any solutions but provide options to the teachers to fulfil their tasks in a more developed way as long as it is in the directions of the overall goals, earlier mentioned.

Another main point is that the management not only put pressure on the organisation to develop. They also provide support in form of creating this development project, the Breakthrough Project that provides new opportunities for teachers and others to develop and learn more about teaching and learning in higher education and also changes their practice. In the project, an extended number of educational consultants work at LTH. Most of them are recruited among teachers at LTH and some are professional educationalists and staff developers.

3.6. Consultative support as a learning process

The main road is not to tell teachers how to solve a problem but provide support and knowledge to make it possible for them to find their own solutions and enriching their own knowledge. From a learning point of view, this way of working with development should lead to more durable changes. Therefore, all development activities must start with the teacher's own interpretation of a situation, description of the problems and formulation of strategies and objectives.

This corresponds to what (Ellström, 1996) labelled learning for development. If a developer interferes too much in this process, there is a risk that the teachers feel that new material is imposed upon him or her. The result is either rejection of the new material or learning for adaptation. The supporting person has to listen carefully, encourage the teacher to describe the situation, and encourage the teacher to relate aspects within the description. In this process, contradictions, problems or tensions will be visible, and if it succeeds they will be apparent for the teacher, even discovered by him or her. The next phase is to ask for possible solutions already in the teachers mind. Once they are described, the support person may encourage the teacher in that direction but also introduce alternatives close to the one the teacher preferred. This is in order to widen the range of choice and to make even more aspects visible to the teacher. In this process of introducing alternative ways to act, the support person may also get in contact to teachers who already have tried similar things. The work of the support person, as described here, resembles Carl Roger's version of a learning facilitator (Rogers & Freiberg, 1994).

Development processes in organisations is similar to learning processes and it is the same principles that is the basis of a good learning outcome and a durable change in developing organisations.

One of Carl Rogers' central conceptions is "significant learning" (Illeris, 1999), which is that learning results in real changes in a person's way of thinking and behaviour. In organisational development, these permanent changes are the main purpose but they are often very difficult to achieve. This is connected to the factor of preventing things from going back to normal mentioned in chapter 3.

This reasoning is connected to Piaget's theory of learning. Piaget, and more with him, thinks that the human being always strives for balance in the meaning and that experienced conflicts, psychological or intellectual, cause stress and anxiety. In some way, this feeling of discomfort has been taken care of. Still, it is probably one of the main reasons that we learn. New things we meet or experience must be dealt with in some way. If the conditions are not optimal, this feeling can raise our psychological defence mechanisms. Central conceptions in Piaget's theory are assimilation and accommodation. Assimilation is when new experiences are linked to already exciting structures of knowledge. This process makes our cognitive schemas more complex. Accommodation is the individual adaptation to the surrounding world. Accommodation takes place when it is no longer possible to insert experiences from the surrounding world in the already existing intellectual structures because the difference is to large. The old structures must be reconstructed and this procedure takes a considerable amount of energy. This is a real reconstruction in the way of thinking, interpreting and viewing the surrounding world. Three important conditions have to be fulfilled to make a durable change. They are the already existing cognitive structures which can be reconstructed, a need and an interest to mobilise the necessary energy and strength and, finally, a sense of security in the situation (Illeris, 1999).

3.7. Changes of attitudes and values

Cultural changes are about changes in attitudes and underlying values among the employees in an organisation. The basis for humans' actions is their personal and internal attitudes and values, which are not always conscious. Attitudes and values are often hard to change because they are a part of the fundament that makes us experience some kind of stability in how we understand and behave in our context. The strategy in this project, to try to achieve a change in the teachers' attitudes and values towards student learning and how teaching could be made in a different, more sufficient way to achieve better results, have been to try to create and find good examples in the teaching activities at LTH. The role of these good examples is to show teachers that it is possible to work in different ways and share the experience of the teachers who develops new ways to teach and handle learning situations. Changing attitudes and underlying values is hard work and take effort and time. Often the ground to these processes is to have experiences of something new and find that this could be positive for ones person (Berg et al., 1992).

In a large-scale project where a majority of the organisation is involved in some way or at least noticing that something is happening, it is also a question of how different persons in the organisation react to these new circumstances. A paradigm shift leads to that what used to be the best way of doing things not is the best way anymore and new ways are encouraged. The result of this is that people successful in the old paradigm maybe not are the leading persons anymore. It is also very important to be aware of these processes in the organisations (Burns, 1984).

The issue of people's self esteem must always be considered. It guides our social life and the way we interact to a very high extent. That's why it is also important to think of those who are against changes and rethinking.

In the Breakthrough project, the strategy is to not criticize this reluctant group too heavily, but to try to listen to their arguments. This has two advantages. Firstly, the persons have an opportunity to discuss their point of view. These discussions can give new insight into people's reactions to the development process. Secondly, to put down people and make them even more reluctant. For some people changes and rethinking take time and, for different reasons, it is for some people not possible at all.

It must be taken in to consideration that a university is a very special kind of organisations with features that to a high extent have an impact on its employees and the psychosocial working environment.

3.8. Levels of responsibility

In an organisation there are different levels of responsibility. In order to get different parts and levels of the organisation to co-operate, it is important that everybody's role is clear and the right level is addressed in certain questions. In the beginning of this project, it was important that the person in charge had the right opportunities and competencies to fulfil his or her role. If there have been problems with this, measures have been taken to improve the situation and to make the areas of responsibility more clear. It is also critical that it is well known to everyone and obvious how the organisation works. In other situations there is a great risk to support the creation of a "hidden organisation". This will create suspicion between the members of the organisation and prevent any effective flow of information. The result is at best quality status quo, but more likely a decline in quality.

The Breakthrough Project rests on the assumption that freedom is only experienced when the frames to act within are well known. Not knowing the frames for one's actions and responsibilities can have different impacts on different persons. For some it leads to that they do not do anything at all and experience that they are not allowed to. Others are doing everything they want and experience no limitations for their actions until they break all rules. It is the responsibility of the management to make this frame for actions at different levels well known. This creates a feeling of certainty and helps everybody to handle their responsibility in appropriate ways.

3.9. Financial issues

The project started with no additional financial resources. All parts are financed within the normal funding of the Institute of Technology. There is a risk that everything done with extra resources would not be considered as "normal". If new activities are planned and tested "within", they will be influenced by the "normal" during their development stages and therefore they will gain a better chance for survival.

The financial resources go to the educational consultants and the different activities in the project. In order to ensure the willingness to spend money on a pedagogical development project, it is important to show the different members of the organisation, departments, teachers and boards of educational programs, that the project supports them and improves their activities. It is also important to work closely together with the members of the organisation, respect their working conditions and to win their confidence.

4. Five main factors to deal with

According to experiences from the Breakthrough Project, five factors have to be dealt with in large-scale development projects.

These can be summarized as:

- Large-scale development projects must deal with the tendency of things to go back to "normal".
- Large-scale development projects must take the dissonance in identity into consideration and be prepared to meet its challenges.
- Large-scale development projects must focus on all groups and levels in order to get them involved.
- Large-scale development projects must address the trap of privacy.
- Large-scale development projects must find ways to develop the necessary leadership to govern the development of quality in teaching.

Four of the factors will be more closely described below. If they are fulfilled, the experience from this project says that the first factor, the tendency of things to go back to "normal", is dealt with as well.

4.1. Matters of identity

Like all professionals, university teachers strive to do their best. As in many other organisations in Sweden, employees at universities have to deal with heavy workloads. For many years, funding from the government has been decreasing and, at the same time, there is a demand for education of more students with more diverse backgrounds. As a consequence, there is a lack of resources and a need to prioritise.

Most teachers at LTH are also researchers. When they do research they use concepts and perspectives and skilfully apply them to the world. They are used to discuss complex questions and to formulate knowledge on advanced levels. However, this is only true in the context of research. Many of the researchers lack words and perspectives when it comes to teaching. For these persons, a change of focus from their research subject to teaching is a matter of identity. It does not only have to do with status, it also has to do with the experience of dealing with problems in a professional way as they do in their research, where they have the knowledge and the skills. Dealing with problems from a different standpoint in a subject they are experts in, teaching, where they often have a severe lack of conscious knowledge can of course be difficult and cause a feeling of insecurity that do not correspond to the most teachers identity as an expert in their field of research. Consequently, teachers/researchers tend to aim for a much higher degree of accuracy when reflecting upon research than they do when reflecting upon teaching.

One of the purposes of the Breakthrough project is to give issues of teaching the same level of status as issues of research. This will also change how careers are made and make the conditions for being successful different. This, of course, has an effect on the whole system and many people have to rethink the way they are used to act. This has consequences for their identity as well.

Large-scale development projects must take this dissonance in identity into consideration and be prepared to meet its challenges.

4.2. All groups involvement

Pedagogical development is not new. Some teachers have always developed their teaching and found ways to develop their competence. In discussions they are often referred to as the already "blessed" or "redeemed". They often consider themselves to be special, as the odd ones, sometimes even as the rebels in a hostile system. Maybe it is now time to make this effort to develop pedagogy to a matter concerning the whole organisation. Probably there is a need to change the traditional way communication is taking place.

In Sweden, there is a long tradition to try to change the pedagogical culture of the universities by supporting the rebels. The council for the re-newel of higher education (http://www.hgur.se/), established in the beginning of the 90s, actually had it as an explicit principle only to fund projects where a single teacher tried to change a specific teaching situation together with his or her students. Organisations and departments could not get funding for development projects. Obviously, the underlying assumption was that the individual teacher within a hostile system was the one that deserved support. To fund departments, for example, were like supporting a structure never able to develop.

There are other teachers, who admire these developers, but they do not do the same, or at least they do not develop their own teaching and competence to the same extent, and they do not take part in discussions about teaching. Another group is the developers opposite. They advocate the system as it is and oppose arguments for change because, as they say, "secure the quality as it is". If this group is active and take part in discussions, there are other teachers favouring them in silence.

1. Favour change and active	3. Favour the present, and silent
2. Favour change and silent	4. Favour the present and active

In some contexts, group 1 and 4 regard group 2 and 3 as the silent majority. The critical aspect in this context is the fact that group 2 and 3 do not take part in discussions. Group number 2, which sympathises with group number 1, which constantly strives to develop the culture of teaching and learning, voted for change but not to take active part in the change process. Group number 3 sympathises with the conservatives, the ones opposing new methods in the teaching activities, but do not actively take part in the debate with group number 1. Group 2 and 3 are quite passive in the open debate. They become the audience watching group 1 and 4 struggling about which paradigm that is right.

It is important to find ways and forums in the academic society where everybody who wants to discuss and express a point of view feel they can do so in spite of academic degree or being a person always talking with the strongest voice. At many universities, this is indeed a cultural change. When people learn they can have a good and constructive dialogue, that makes them express their thoughts and learn that other are listening, development have a good opportunity to begin.

In the Breakthrough project the key to a durable change in the learning community at LTH is to make group 2 and 3 actively involved in debate on pedagogical issues. Their pedagogical competence is the critical aspect. The assumption in the project is that if these groups start to discuss teaching and learning with each other in a professional way, they will start to challenge each other's perspectives, strategies and interpretations and start to co-operate. They will do this because they work in a scientific context. The idea that the only way to secure quality when dealing with unknown matters and knowledge is to ground it in a discussion with colleagues and literature is so deeply ingrained in a scientific community that it is an absolute imperative for everyone to engage.

The result is a Community of Practice that (Wenger, 1999) focuses on as a research field. According to Wenger, it leads to constant learning and development. If it would be possible to start a process where teaching and learning is dealt with in a similar way, the development of a new practice would start without external support. The result would be an imperative to treat teaching scholarly.

Large-scale development projects must focus on group 2 and 3 in order to get them involved.

4.3. Teaching as a privacy matter

Traditionally, most teaching at LTH has been a kind of private activity for the teachers. Often a load of teaching is allocated to one teacher. If there are a large amount of students at a course, more teachers, postgraduate students, other teachers and senior students, are often linked to the course. The responsibly teacher is responsible for this team. In the most departments there is no tradition to pay attention to the on-going teaching as long as the passing rate is fairly normal, between 50% and 75%. There is no tradition to ask about gained experience, documentation, a description of the teachers' pedagogical philosophy or the idea with the pedagogical design.

The result, as it has been described, is that either the teachers in charge develop exclusive and elaborated teaching strategies on their own, often with a high dependency on their own effort and energy; or they do not do anything except what is considered to be normal. The most obvious weakness in this situation is that the teachers and other actors do not learn from each other because there are no forums where they can discuss teaching and no cultural impacts to do so. There are no available documents about other teachers´ reflections upon teaching experiences. It is also very difficult for leaders and heads of departments to govern the activities concerning teaching and allocate the right resources to different courses because they lack information about their teachers´ skills and about the learning process in courses.

A community of practice is characterised by a number of people united in a *mutual engagement* for *a joint enterprise*, and while pursuing this they develop a *shared repertoire* (Wenger, 1999). In the case of LTH, this means that teachers come to regard good teaching and its purpose, student learning, as a shared enterprise in which they can engage and draw on each other's experiences. It will include reading about teaching and learning, participation in pedagogical courses, writing about teaching and learning, the engagement in different development activities, and so on. The intended outcome is the development of a culture where teaching and learning is discussed by different teachers and between different levels within the organisation.

Large-scale development projects must address this trap of privacy.

4.4. Issues of leadership

Questions of leadership are always central in any organisation. At LTH, the important levels are the Dean's level, the heads of the Department and the boards governing the educational programs. The traditional academic leadership is the collegial. In some ways, it can be described as a weak kind of leadership. In other ways, it can be described as an extremely strong one. It is mostly built to support the process of quality control, in many ways unique for the academic society. The object for this quality control is knowledge, new unique forms of knowledge. The ideal and often mystified form it takes is the discussion between peers. Only in a qualified discussion can new knowledge be tested and validated. A collegial leadership supports this. However, it is not constructed to initiate and support change, it is not even constructed to ensure and develop quality in teaching. Today, a study program at LTH involves several hundreds of people, students, teachers and support staff.

Today, universities are put under pressure from many directions: more students with more diverse backgrounds, the government demanding proof of accountability, students demanding value for money and up to date pedagogy, and so on. Under these circumstances, good leadership is absolutely vital. There can be no possibility for individual teachers/researchers to focus both on their actual activity and excel in this and at the same time be oriented in the context of a whole university and its relations. Trust must be cultivated in a culture of critical thinking, individuality and competition.

In a 1999 survey, teachers teaching in the Computing engineering program expressed confusion over what was pedagogically expected of them. They did not know if they were allowed to or even expected to experiment in their teaching. They felt isolated and without support. In a traditional academic sense, they were absolutely free to do whatever they wanted as long as the pass rate in the exams was fairly stable and there were no student complaints. However, this perception of freedom just do not match today's situation. When put under pressure, these teachers do not have much to do but to work harder, using the same strategies more intensely.

At the beginning of the Breakthrough project, it was important to establish a sense of cooperation between the teachers. To do this there has to be a discourse focusing on teaching and learning and a common language. Teachers and others at LTH have to talk to each other and listen to each other's descriptions and slowly recognise similarities between different contexts. In the project, the pedagogical courses, the consultative support to individuals and groups, and the campus conference are the main measures taken in order to address this issue.

Once a shared view of the teaching and learning situation within the faculty is constructed, it will be apparent for the teachers that certain problems are more efficiently dealt with on department level or even on faculty level. Leadership is constructed in the relation between the lead and the leaders as a result of meaningful interaction. This way of reasoning is grounded in a social psychological perspective (Berger & Luckmann, 1966; Asplund, 1987) where reality is constructed within the process of interaction and based on interpretations of actions. With the perspective outlined, it follows that leadership is a relation based on a common shared understanding of what matters in the organisation. This is fulfilled in the project by building this understanding from the bottom, letting the actors develop their descriptions in interaction with each other until a common understanding is constructed. It is also crucial that the management level are clear about what they want the organisation to achieve and why.

The process is grounded in literature concerning conceptual change (Posner, Strike et al., 1982; Demastes, Good et al., 1995) In short, a person who is going through a conceptual change or a conceptual development has to experience the old conception as problematic. For a new perspective or conception to be established, the person must regard it as fruitful. It has to solve the old one's problem but also seem promising to explore further. If these and other conditions are met, a conceptual change is likely to occur. If managers at universities demand change from teachers who are not experiencing any problems with older ways of doing things, the teachers are more likely to treat the managers and their demands as the problem (Trowler, 1998). As a result, the teachers will pretend to do as the managers want, in order to release the pressure, but in reality nothing will change.

Large-scale development projects must find ways to develop the necessary leadership to govern the development of quality in teaching.

5. Examples of activities in the Breakthrough project

To reach the goals of this large-scale project, activities that support the development of the employees and give them the above-mentioned experiences of new ways of acting and thinking in teaching and learning situations must be carried out. It is also important to give the teachers and other at LTH several opportunities to rethink processes concerning student learning and the teachers' situation and how they can be handled in ways that better support student learning. Examples of such activities in the Breakthrough project are described below.

Pedagogical competence development
According to (Ellström, 1996) learning might be adaptive or developing. Learning for development takes place when the learner has as much control over the process as possible. Critical aspects are *the objective, the method* and *the assessment.* If the learner has control, the learning is likely to lead to development. The designs of the pedagogical courses are influenced by these principles.

Several courses in teaching and learning are offered to the teachers at LTH.

Seminars
Seminars are offered in areas of special interest and novelty. They are also opportunities for discussions among the teachers.

Campus conference
A campus conference about teaching and learning is arranged annually. Teachers at LTH are welcome to send in abstracts about issues of teaching and learning. The abstracts are often about try-outs they have had in their teaching and the results of them and other questions of concern.

Team-planning
The boards of the educational programs have decided to organise the teachers in teams to take responsibility for curriculum planning and teaching that involves more than one department. To support this processes, consultants working in the Breakthrough project offer support to the teams.

Educational consultants
An extended number of educational consultants are employed at LTH. They offer support in questions of teaching and learning to everyone who asks for it. They are also responsible for the teaching in and organising the courses in teaching and learning at LTH. The team working with the Breakthrough project consists of teachers at LTH and staff developers with competencies in teaching and learning. This leads to a good mixture of practical based and theoretical based competencies.

The Pedagogical Academy
This is a way to reward professional teachers and their departments. Teaching portfolios and interviews are used to distinguish these teachers. They will, besides from a raise in salary, earn the title ETP (Excellent Teaching Practice) and as such stay in the system as a long-term investment in relation to the ongoing discussion about teaching and learning within LTH. The Pedagogical Academy is described in another article in this publication.

6. Present state of the Breakthrough project

From the first of January 2004, The Breakthrough project is a permanent activity at LTH. The dean at LTH is still leading the program and operative responsible is the director of the human resource department and the director for the Breakthrough program. It is now part of the Human Resource Management function at LTH and it is linked to other competence development programs. The strategies are still the same as described above. Nothing is changed in the grounded philosophy about how to implement changes in the teaching and learning culture at an institute of technology.

During the last year of the project phase, the Breakthrough project was evaluated by Learning Lab Denmark in co-operation with Centre for Higher

Education, Greater Copenhagen and by Learning Lund, centre for research on teaching and learning in higher education, Lund University. Reports from these evaluations are expected during 2004. From already known results, actions have been taken. The Pedagogical academy is paused this year to be further developed. To strengthen the striving towards a culture of Scholarship of teaching, research and competence development in didactics are right now being implemented at LTH.

Research on teaching and learning activities at LTH is starting in 2004 as a new activity in the Breakthrough program. In this process, researchers from Lund Institute of Pedagogic will work together with the consultants in the Breakthrough project. The first project is about the features of examination of students at LTH. External funding will be applied for to finance this research.

From 2003, it is compulsory for all new lecturers in Sweden to have formal education in teaching and learning in higher education. At Lund University, it has been decided that all teachers at the university shall have ten weeks (400 hours) education in teaching and learning. New employees are supposed to have done five weeks courses in teaching and learning before employment and within two years they shall take five more weeks. This have somehow changed the circumstances for the Breakthrough program and raised new questions to deal with.

All this is about creating a suitable and good learning environment for the students, teachers and others within and in contact with LTH.

References

Asplund, J. (1987) *Det sociala livets elementära forme*r, Göteborg, Bokfölaget Korpen.

Barr, R., B. & Tagg, J. (1995) From Teaching to Learning – A New Paradigm for Undergraduate Education, *Change* (Nov/Dec): 13 - 25.

Berg L-E et al. (1982) *Medvetandets sociologi - en introduktion till social interaktionism*, Whalström och Widstrand, Stockholm.

Berger, P. & Luckmann, T. (1966) *The Social Construction of Reality. A Treatise i the Sociology of Knowledge*, Penguin Books.

Bowden, J. & Marton, F. (1999) *The University of Learning*, Kogan Page, London.

Boyer, E. L (1990) *Scholarship Reconsidered.* Priorities of the professoriate, The Carnegie Foundation, New Jersey.

Burns R. B. (1984) *The Self concept, theory, measurement, development and behaviour*, Longman group LTD, Singapore.

Demastes, S., Good, R. et al. (1995) Students' conceptual ecologies and the process of conceptual change in evolution. *Sience Education,* Vol. **79**: pp 637 - 66.

Egidius, H. (2002) *Pedagogik för 2000-talet*, Natur och kultur, Stockholm.

Ellström, P.-E. (1996) Rutin och reflektion. Förutsättningar och hinder för lärande i dagligt arbete.In: *Livslångt Lärande.* P.-E. Ellström, B. Gustavsson and S. Larsson, Studentlitteratur: pp142 -179, Lund.

Fromm, E. (2000) *The Art of Listening*,Natur och kultur, Stockholm.

Healey, M. (2000). Developing the Scholarship of Teaching in Higher Education: a discipline-based approach. In: *Higher Education Research & Development* Vol. 19, No. (2): pp 169-189.

Illeris, K. (1999) *Lärande i mötet mellan Piaget*, Freud och Marx, Studentlitteratur, Lund.

Kolb, D., A. (1984) *Experiential Learning. Experience as The Source of Learning and Development*, Prentice-Hall, New Jersey.

Kreber, C. (2000) How University Teaching Award Winners Conceptualise Academic Work: some further thought on the meaning of scholarship.In: *Teaching in Higher Education,* Vol. 5, No. 1, pp 61 - 78.

Kreber, C. (2002) Teaching Excellence, Teaching Expertise, and the Scholarship of Teaching. In: *Innovative Higher Education,* Vol. 27, No. 1, pp 5 - 23.

Kuhn, T., S (1992) *De vetenskapliga revolutionernas struktur,* (The Structure of Scientific Revolutions, 1970), Thales.

Posner, G. J., Strike K. A., et al. (1982) Accommodation of a Scientific Conception: Towards a Theory of Conceptual Change. In: *Science Education,* Vol. 66, No. 2, pp 211-227.

Prosser, M. & Trigwell, K. (1999) *Understanding Learning and Teaching.* The experience in Higher Education, The Society for Research into Higher Education & Open University Press.

Rogers, C. & Freiberg, H. J. (1994) *Freedom to Learn*, Prentice Hall Inc.

Trigwell, K., Martin, E. et al. (2000) Scholarship of Teaching: a model. In: *Higher Education Research and Development*, Vol. 19, No. 2.

Trowler, P. (1998) *Academics Responding to Change.* New Higher Education Frameworks and Academic Cultures, The Society for Research into Higher Education & Open University Press.

Wenger, E. (1999) *Communities of Practice.* Learning, Meaning, and Identity, Cambridge University Press, Cambridge.

Chapter 5

Creating a learning environment for engineering education

Hans Peter Christensen

1. Introduction

Until recently discussions about improvement of educational quality have focussed on the teacher – it was assumed that by training the teacher you could increase the students' learning outcome. Realising that other changes than better teaching were necessary to give the students more useful competencies, the idea of faculty development was introduced. But even this is not enough. For some time it has been said: 'from teaching to learning', but very little attention has actually been on the students' learning through active studying; how should the student work in a learning-effective way? And the introduction of IT has highlighted the importance of the learning environment, but the focus has narrowly been on the physical environment. However, the mental framework is also very important. To assure educational quality it is necessary to take all these elements into account and consider the total learning environment as an integrated whole.

2. General description of a learning environment

Just like teaching requires three elements: a student, a teacher and a topic to be learned, a teacher-facilitated learning process has three inputs: *studying*, *facilitating* and *framing*. As the topic to be learned sets the intrinsic conditions for teaching, the framing sets the external conditions for learning.

The *learner* is playing a double role central to the learning process. The output is to the learner, and as a student the learner supplies the studying input.

This feedback loop is essential to the learning process, since new learning builds on existing knowledge (Bransford, et al. 2000). The *teacher* supplies the facilitating, and the framing is given by the *context*, which includes physical surroundings as well as mental conditions.

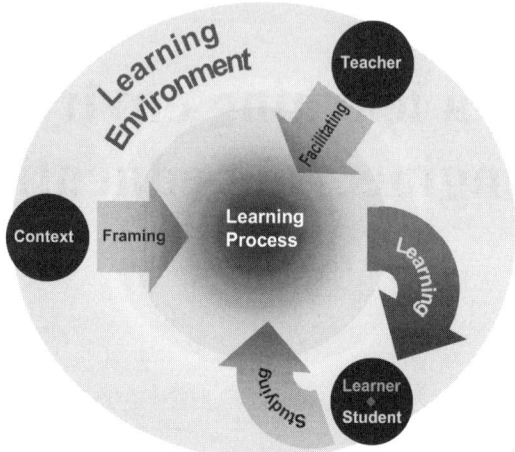

The three actors in the learning environment are depending on each other in giving input to the learning process. The framing sets limits for the possible teacher and student action – but the student and teacher are part of the learning context, so their actions influence the framing. The facilitating encourages certain kind of actions from the student, but the teacher's options are influenced by the student response. For these reasons it is not sufficient to discuss the elements of the learning environment independent. The output from the learning process is very dependent on how the actors in the learning environment interact.

2.1. An institutionalised university learning environment

The objective for a formal learning environment is to support the learner in learning something – to optimise the learning process. In the formal learning environment the framing is supplied by the institution, and the framework consists of the teaching facilities and the agreed rules and policies for the educations as well as informal traditions. The framework may not be the most important factor in the learning environment, but certainly the most basic factor in determining the quality of the learning obtainable. It can be flexible designed, but in the end it always put some limits on the student's way of studying and the teacher's way of teaching.

Since learning is an active process, you have to work to learn. So studying means working, following that a learning environment should be a laboratory. *'Learning for understanding occurs principally through reflective engagement in approachable understanding performances'* (Wiske, M. D., 1998) – i.e. deep learning is obtained by active doing. Activity does not necessary mean physical activity; activity can also be mental. However mental activity is not suitable for all kind of learning. Procedural knowledge is learned by physical activity (Bech, 2003).

Prescribed learning environments operate in two modes — *formal* and *informal* (Crawley, Hallan & Imrich, 2002). So even though a formal learning environment is designed to operate in the formal mode, its informal operation should also be taken seriously into consideration. Actual, in an active learning environment the formal and informal modes could be difficult to separate, since many formal as well as informal activities will be based on independent work individually or in groups.

2.2. The double-loop feedback process

The feedback loop in the learning process is actually a double loop. On basis of the experimental learning circle (Kolb, 1984) the following model can be proposed.

The student does (on basis of his previous knowledge) some active *experimenting* as input to the learning process, and as learning output the student gains *experience*. The student now *reflects* on this experience, and this input hopefully results in a learning output in form of conceptual *understanding*. (The difference between experience and understanding can with reference to (Aristotle, 350 B.C.) be explained as follows: Experience is knowledge of the particular; understanding is knowledge of the universal).

On basis of this new understanding the student can do new investigations. The first loop could be called **discovery** and the second **conceptualisation**. The two loops are also related to Piaget's ideas on assimilative and accommodative learning (Illeris, 2002) – first you try to fit new knowledge into your current understanding, but as new knowledge accumulates you may have to revise your models bringing new understanding. The greatest danger in active learning is to put too little attention on the second loop (reducing the learning cycle to the engineering cycle: Try something; if it doesn't work try something else!).

The learning environment must support this double feedback loop. To support the first loop the learning environment must encourage activity. The teacher input must support the initiative of the student and not unnecessarily control or limit his behaviour. And the framing should define a very flexible framework that leaves room for experimenting. The second loop will be supported primarily by input from the teacher. 'From teaching to learning' does not mean that the teacher *not* should play an important role in the learning process. The teacher should not take over the conceptualisation process, but she should certainly make sure that the student does not end up in the engineering circle. But the framework should also support conceptualisation by not overloading the curriculum, but giving time to absorption in the topics.

3. A learning environment for engineering competencies

The design of a learning environment depends on the competencies the learning should give the student. In engineering education the learning should result in engineering competencies.

3.1. Engineering competencies

Competencies can be defined in many ways. Very often it includes the following three elements:
Knowledge
Skills
Attitudes
But competency is not just the combination of these three elements. Professional competency must include the ability to use these elements.
A definition of **professional competency** could be:
The ability in a given context to apply knowledge, skills and attitudes to analyse and solve authentic problems.

Knowledge and skills can be obtained at different levels: Knowledge goes from knowing facts to having deep conceptual understanding. Skills go from being able to follow a script or procedure to complex problem solving. Professional competency requires high levels of knowledge and skills.

Skills and attitudes come in two varieties. A skill can be technical or personal, and both are of equal importance at high skill levels – real-world problem solving requires very well developed personal competencies. An attitude can be professional or personal; as a professional there are certain standards in ethics and method you have to follow, and in a given job you need certain personal qualities.

Engineering competencies could then be defined in the following way:
The ability to
handle complex situations in a way, which on one hand includes the use of scientific methods and on the other meets demands and expectations from employers and users.
analyse ill-defined open problems and if relevant reduce these to solvable technical problems.
evaluate, select and optimise a useful method for solving a problem, elaborate a solution, and evaluate the quality and limitations of this.

It is clear from this definition that engineering competencies includes a lot of *tacit competencies*. It is not possible to give the student all these competencies during their formal education, but the learning environment should make it possible to achieve the foundation for these competencies, so the graduate is well prepared for his future life-long learning of these tacit competencies.

This implies that the learning environment should be designed, so that it does not only support giving the student deep understanding, but also supports giving the student technical as well as personal *problem-solving skills* and *learning skills*. It must encourage engagement in **active learning** in the form of hands-on authentic problem solving as well as **cooperative learning** to develop personal competencies.

This puts strong requirements on the input to the learning environment from the perspectives of all the actors.

3.2. A learning environment from the student's perspective

Studying is working, but the learning outcome is properly more dependent on the quality of the study work than on the quantity. Generally the student can take two different approaches to learning – a *surface approach* and a *deep approach* (Ramsden, 1992). In the deep approach he really tries to figure out what it is all about – to understand and seek a meaning. The focus is on 'what is signified'. In the surface approach he just tries to get a good grade by reproduction and rote memorising. The focus is on 'the signs'.

Since the deep approach leads to understanding and transferable competencies, the student must adapt a deep approach to studying. Study strategies with different study activities can lead to this approach, but independent studying and time to absorption in the topic are required. The learning environment must encourage and support study activities compatible with a deep approach.

3.3. A learning environment from the teacher's perspective

To facilitate the student's learning the teacher can take on many different roles – she can be a mentor, a facilitator, a specialist or a didactic professional. The teacher cannot always have the same role; the suitable role depends on the actual situation. But as the student has two approaches to learning, the teacher can have two fundamentally different approaches to teaching described as *teacher centred focus* and *student centred focus* (Prosser,& Trigwell, 1999). In the first approach the teacher thinks the most important thing about teaching is how the teacher performs – that by perfecting this performance she can *transfer* her knowledge to the student. In the second approach the teacher thinks the most important thing is to set up the best conditions for letting the student do the work.

Since the student only can learn by doing the work himself, the teacher should adapt the student centred focus. So the learning environment should encourage and support this focus. It is important that there is *alignment* between the teaching method used and the study strategy applied by the student. Empirical evidence also suggests that the student centred focus help the student to take a deep learning approach (Prosser, & Trigwell, 1999).

3.4. A learning environment from the institution's perspective

The framework includes walls, furniture and equipment as well as formal rules and regulations. But also the informal structures in the form of implicit rules and traditions are important. Learning outcome is very dependent on the context in which the learning takes place (Bradsford, et al., 2002). This means that the framework must be carefully designed with respect to the desired competencies of the graduates.

The **physical surroundings** must be designed for student activity individually and in groups.

The way a campus is planned and the buildings structured have much more influence on the possible teaching methodology than normally acknowledged; teaching methods are built in bricks and concrete! A campus designed for traditional lecturing is difficult to convert to a teaching method designed primarily around group work. And a campus built with group rooms for project work is useless, if you for some reason have to do large-class lecturing.

Most universities are stuck with the buildings they have, so teachers very often use the framework as excuses for not changing. However, in most cases these objections are only excuses – it is possible to do a lot within existing frames, even if it is not possible completely to change the overall teaching strategy.

The ideal physical learning environment would be a very flexible area, where all learning activities could take place. It should be possible instantly to switch from traditional lecturing, to multimedia presentations, to experimental demonstrations, to dialog based teaching (discussions), and - without loosing contact to the teacher - to work in small groups, to work with computers individually or in groups, to do practical exercises, etc. But just as important it should also be possible to do projects and group work in larger groups for prolonged periods of time.

The physical environment for engineering education should be a *real* laboratory. Engineering science is built on theory *and* practice, it cries for physical activity! For some kind of engineering - e.g. electrical engineering - it is easy to work with authentic constructions, but for others like civil engineering it could be more difficult – full-scale bridges are not easy to build... The kind of possible laboratory therefore depends on the field of engineering. A learning environment for mechanical engineering could very well be designed around a mechanical workshop.

Even if the final solution to this problem is yet to be found, there are many suggestions to physically environments. From a MATH-auditorium with IT facilities at the Chemistry Department at Chalmers University of Technology (Chemistry Department of Chalmers University of Technology) to a total integrated learning environment designed around the CDIO (Conceive-Design-Implement-Organise) -concept at the Department for Aerospace and Astronautics at MIT.

Technical facilities include both the instructional equipment (IT etc.) as well as the experimental apparatus for student exercises and projects.

With respect to *IT* the challenge is: How to integrate the students' laptops in the learning environment? How it is possible to integrate many different kinds of laptops in the university's network, and make sure that all relevant programs can run on all computers? How should the institution support the students' laptops – if it should?

With respect to *experimental apparatus* the choices are: Do you want lots of cheap equipment or a limited amount of very good equipment? Do you want inexpensive workshops where students can be in control or advanced laboratories, where students under close guidance are allowed to use expensive state-of-the-art research equipment?

From the above description of the physical surroundings, it should be clear that the right answer from a pedagogical point of view to the apparatus question is: Lots of inexpensive moveable equipment, which the student can control – i.e. equipment that only is used by one group during a semester, a year or maybe a whole study programme. Of course engineering students cannot complete an engineering education without contact to very specialised and expensive apparatus, but for most practical work general inexpensive equipment will do – certainly the first years of the education. The era of labs with fixed set ups for cookbook exercises is over!

Rules and regulations – i.e. structures of study programmes, semester layout and timetables, and rules for assessment and evaluation etc. significantly influence the learning environment. Rigid rules limit the learning outcome.

A study programme should be structured, so it gives the student time for absorption in the topics – i.e. curriculum overloading should be avoided. Time schedules should be so flexible that it leaves room for the teacher to use all kind of teaching techniques. The rules for assessment should be designed, so they don't force the student to a surface approach; it should be possible to evaluate understanding and competencies.

3.5. The interaction between the student and the framework

In order for a learning environment to be of use, the learner should physically and mentally be in it! The student should feel that it is advantageous for him to be there. The environment should make the student feel attached to the environment – he should feel that he belonged there, and not just a place where he was invited to come and spend some hours. This means that each student should have a place of his own – e.g. some kind of locker. But since most teaching includes some kind of group work, there should also be a place for each group, where it would be possible to create the group's own environment. (This would probably cause a lot of resistance from teachers, who think a learning environment should look tidy!) If we want students to work actively – and not just sit reading at the library – we must provide them with these facilities.

4. A meta-learning environment

Meta-learning is learning about the learning process itself. It is not what to learn, but how to learn it. To learn you have to know how to study, so a meta-learning environment deals with how to make the student a good learner.

In order to adopt a suitable study strategy the student must be reflective on his situation. He needs to focus on what the learning objectives are and how to reach these objectives under the given circumstances. The student should know why different teaching methods are used in order to respond properly – especially for teaching methods requiring the student to take initiative on his own.

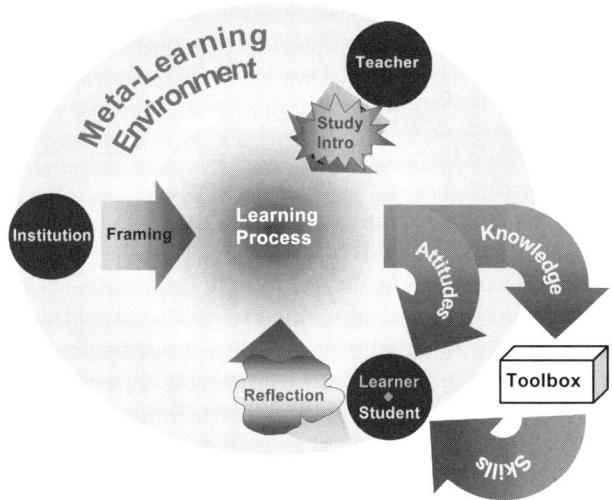

To do this in a qualified way the student needs an introduction to the used teaching methods and given framework. (The general study introduction should be supplemented by clear statements and discussions about goals and methods in each course.) The student needs a toolbox with the necessary tools for making choices and tools for surviving and learning with efficiency, where these choices lead him. Students are not born with a study toolbox, so the student must be given a box and opportunities to fill it up with useful tools. He should be professionalized as a learner.

4.1. Introduction to the study for new students

The general objectives for introduction programmes for new students are to facilitate the students' transition from previous teaching methods to those used at the university, introduce the educational programmes, and integrate the students socially in the local environment. In the following will be focussed on the student's response to university teaching, although social integration also is very important for a learning environment.

There are different strategies for study introduction (Christensen, 2000):

The tool/skills approach: You can give the students some skills (tools and techniques) that they immediately can apply to their studying.

The enlightenment approach: You can tell the students how you learn, what the intentions of the educational programmes are, why certain teaching methods are used, and how the teachers expect the students to react.

The first approach is often used and the introduction the students expect. But it is very doubtful, if this method gives the expected results; the students may learn some good tools and techniques, but if they don't know why they are using these, they may end up being very efficient with respect to studying but not at all with respect to learning. It is necessary that the students develop critical reflection about what they are doing. The focus should be shifted to a discussion on how to obtain deep conceptual learning and useful competencies.

The second approach might therefore be the right basic principle for an introduction, if deep learning is to be promoted instead of surface learning. The enlightenment should be complimented by offers of introduction to tools and techniques, but the students have to realise a need to change habits before they can appreciate new tools and techniques. The introduction should encourage students to change from passive behaviour to active learning, but agreement between the promoted study methods and the actual teaching methods must be assured.

Skills at the operational level should be trained in the courses, where they will be used – you cannot train skills without using them on something realistic. IT skills should be introduced to the students the first time they log on to the university network or are required to do a computer modelling. Teamworking skills the first time they meet group work. Working with projects the first time they have to do project work. Report writing the first time they have to write a real report. Presentation techniques the first time they have to make a presentation. Information search first time they have to find important literature. Traditional study techniques (note taking etc.) should be offered to the students who feel a need for it and when they feel this need.

5. The extended learning environment

The primary learner in a learning environment is the student. But he should not be the only learner. In order to develop her teaching, the teacher should also be a continuous learner – and for this there should be a local *teacher learning environment*. And in order to develop the framework, the faculty as a whole should be in a state of learning – so there should also be a local *faculty learning environment*.

5.1. A teacher learning environment

In the teacher learning environment the teacher is the learner studying primarily by reflecting on her teaching and the learning it induces. The teaching function is performed by the teacher-training personnel in supporting the teacher in developing her teaching. The local framework is formally set up by those deciding staff policies, and informally by the cultural traditions in the faculty, department or teaching group.

In order to adopt a suitable teaching methodology, the teacher must reflect on what she is teaching and why she is teaching the way she is (Walkinton, Christensen & Kock, 2001). She needs knowledge about learning processes and the students' prior knowledge. The teacher needs a toolbox with different teaching and assessing methods, so she knows what options she has, when she wants to give the student a certain competency. But it is not enough to know the methods, she must be able thoroughly to analyse the methods with respect to different quality parameters and outcome.

To be able to this the university teacher must be given a pedagogical education. This should not be the traditional teacher training supplying young teachers with tricks and quick fixes, but training aimed at a deeper level. It is empirically documented that teacher education changes the teacher behaviour in positive ways (Gibbs, & Coffey, 2001). But it is also clear that teacher training should be a life-long process, involving the teacher in continuous teaching development.

60 · *Creating a learning environment for engineering education*

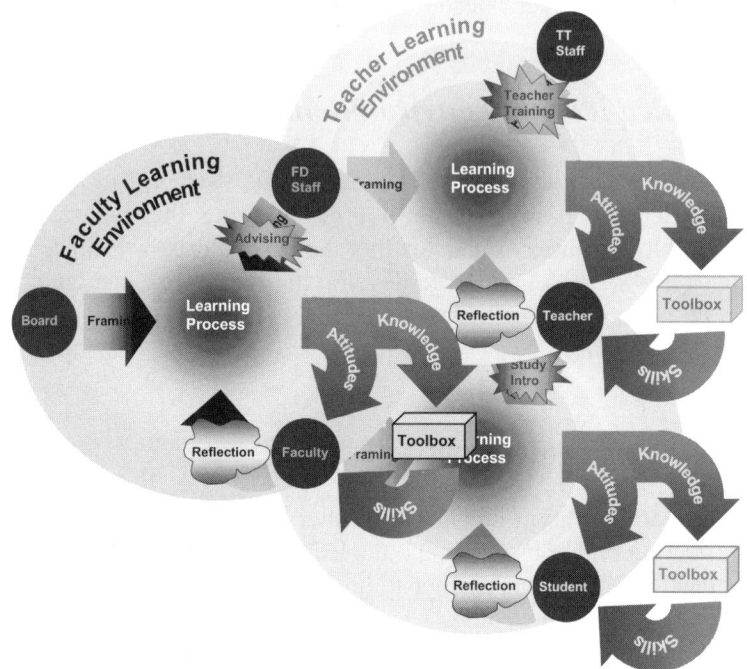

In order for a faculty member to develop as a teacher, she must be professionalized as a teacher. Very often the practice of teaching is implicit for university teachers; they do not have clear explanations for why they teach the way they do (Tenhula, 1996). However, through reflection on teaching experiences it should be possible to clarify one's philosophy of learning and teaching, and by teaching experiments and feedback from students and colleagues it should be possible to find one's own teaching style. In this way the teacher will develop an explicit (conscious) teaching practice and obtain a professional identity as a teacher.

5.2. Pedagogical education for teachers

College teaching may be the only skilled profession for which no preparation or training is provided or required. You get a Ph.D., join a faculty, they show you your office, and then tell you "By the way, you're teaching 205 next semester. See you later." The result is the consistent use of teaching techniques that have repeatedly been shown to be ineffective at promoting learning. (Felder, 2004)

The ultimate objective for teacher education must be increased student learning, but there can be formulated several more specific goals for teacher education, and therefore there also are several approaches to teacher education (Gibbs, & Coffey, 2000):

Behavioural change: The goal is to develop teaching skills and competence. This will typical be done by use of video sessions to develop specific classroom skills in formal training programmes.

Reflective practice: The focus is on responsiveness, student differences, freedom from habits and ability to innovate. This method is normally based on reflective journals and action research.

Conceptual change: The idea is to go from content to process – i.e. to focus on learning; or to put it in another way: From teacher focus to student focus – i.e. to focus on student outcome rather than teacher input. The basis for this is studying learning theory and analysis of teaching methodology.

Teacher confidence: To give the teacher confidence in her ability to teach effectively and to dare to use new methods. Strong support, positive feedback and a willingness to accept failures promote this.

All these aspects are important to teacher education, but too much focus on teaching skills is very conservative, so if you really want changes, behavioural change should not have top priority – it could be argued that it should have no priority at all in university teaching. Reflective practice is more important than formal abilities. Focus should be on the three other goals with top priority to conceptual change, since this is most directly linked to the overall objective of increased student learning; reflection and confidence could be considered means to make conceptual change possible.

A teacher-training programme can consist of lectures, seminars and workshops, but it could also be practical on-the-job training supervised by a well-respected older teacher – and very often it will be a combination. This raises the question on how to balance a theoretical approach versus a practical approach. It is often argued that you cannot teach how to teach; therefore the main emphasis should be on mentoring learning-by-doing. But the situation is not that simple.

It is probably correct that you cannot teach how to teach; teaching is a far too complex activity to learn in the classroom. Effective teaching depends on many factors – the teacher, the students, the subject and tacit competencies, which the teacher must learn through the procedural system – i.e. knowledge-in-action learned by reflection-in-action (Schön, 1987). But you also need some basic knowledge about learning theory – if the teacher has to figure everything out for herself by inductive learning, it may take a lifetime.

And you need to know what options you have in a given teaching situation – research shows that some methods are better than others for given purposes; neither you have to figure this out for yourself. So some kind of theoretical approach is valuable.

Experience also shows that pedagogical supervision of young teachers is a very difficult business. You need experienced teachers interested and qualified for this job, and both qualifications are in very scare supply. So even when supervision takes place as planned, the value is very often questionable, if the consequences are not directly contra productive in situations, where the teaching quality needs to be improved. Mentoring should be a supplement to something else – even though it is an important and necessary contribution.

The theoretical foundation could be given via an empirical approach based on assignment instead of lectures. It could be inquiries into the students' prior knowledge or conceptual understanding. Or it could be introduction of alternative teaching methods and investigations of these methods usefulness. For university teachers this has the advantage of being research like.

An important supplement to supervision is colleaguevision – i.e. discussion of teaching in small groups of teachers (Lauvås, & Rump, 2001). Colleaguevision can be very successful if introduced properly.

5.3. A faculty learning environment

The faculty learning environment is very important, since it governs the frameworks for the other environments. In the faculty learning environment all faculty members are the learners. The faculty-development people take on the role of the teacher – not by teaching but by advising the leadership. The governing boards of the university deliver the framework, which consists of rules, set up by the institution in accordance with international and national regulation. Official policies may give guidelines for what should be done, but traditions and available resources may in the end determine what in practice is possible. The culture of the teacher's surrounding is very important; it can encourage spending time on teaching development, and - very often unfortunately - it can hinder motivated teachers in doing, what they think they ought to do and would like to.

What the faculty should learn is how to support effective learning and teaching, and how to develop student and teacher learning environments. Primarily faculty members should learn how to create a positive atmosphere for teaching and learning, how to break down old traditions hindering development of more efficient teaching, and how to reward teachers obtaining good student learning. In other word, learning here is very much about a *mental change*. For this to succeed the decision-makers need professional didactic advises on how to design a fruitful framework. But they also need a toolbox with methods on how to evaluate and promote good teaching (Gibbs, & Habeshaw, 2002).

5.4 Faculty development

Faculty development focuses on *all* employed at a university from technical support staff to the board of directors. It includes teacher education, but it addresses all aspects relevant for student learning incl. advising the governing board on rules, educational management, study planning, staff policies, physical framework etc. Faculty development concerns all elements of the extended learning environment.

With respect to teacher education faculty development can be implemented in different ways (Colet, 2002):

The prescriptive approach: Aiming at teaching effectiveness. Focuses on developing efficient training programmes with formal certification compulsory for young faculty. The teacher-training staff consists of professional teachers qualified in university pedagogy.

The interactive formative approach: Advocates a wide and adapted selection of activities incl. workshops, counselling, formative evaluation and mentoring in order to help faculty develop their teaching skills. Faculty developers are professionals qualified in university pedagogy and counselling.

The distributed formative model: Based on working in a learning community. Supports R&D in the field of teaching and learning. Faculty developers are academics.
From the above discussion it should be clear that faculty development should go to the formative level.

References

Aristotle (350 B. C.) *Metaphysics*, 981 a 15.

Bech, N. I. (2003) *Kroppen skal bruges ved indlæring (You have to use your body in learning),* 'Ingeniøren' No. 7, February 14, Section 2 pp 1.

Bransford, J. D., et al. (eds.) (2000) *How people learn – Brain, Mind, Experience, and School*, Expanded Edition, , National Academy Press.

Chemistry Department of Chalmers University of Technology. Retrieved June 30, 2004 from http://www.cre.chalmers.se/integrationsprojekt/undlok.htm.

Christensen, H.P. (2000) *What students should know to become good learners*, 4th Baltic Region Seminar on Engineering Education, Copenhagen, Seminar Proceedings, pp 27-30.

Colet, N. R. (2002) *Faculty Development Strategies*, Issue Paper, EU, STRATA-ETAN, Foresight for Higher Education/Research Relations.

Crawley, Hallan & Imrich (2002) *Engineering the engineering learning environment*, SEFI Annual Conference, Firenze.

Felder R. homepage. Retrieved June 30, 2004 from http://www.ncsu.edu/effective_teaching.

Gibbs, G. & Coffey, M. (2000) *What is training of university teachers attempting to achieve, and how could we tell if it makes any difference?* International Consortium for Educational Development Conference, University of Bielefeld.

Gibbs, G. & Coffey, M. (2001) *The impact of training on university teachers' approaches to teaching and on the way the students learn*, Early 9th European Conference, Fribourg.

Gibbs, G. & Habeshaw, T. (2002) *Recognising and Rewarding Excellent Teaching – a guide to good practice*, TQEF National Co-ordination Team, CHEP, The Open University.

Illeris K. (2002) *The Three Dimensions of Learning*, Roskilde University Press, pp 27.

Kolb, D. A. (1984) *Experimental Learning*, Prentice-Hall.

Lauvås, P. & Rump, C. (2001) *Vor fælles viden – kollegavejledning som metode til udvikling af undervisning ved højere læreanstalter (Our common knowledge – colleaguevision as a method for developing university teaching)*, Samfundslitteratur.

Prosser, M. & Trigwell, K. (1999) *Understanding Learning and Teaching – The experience in Higher Education*, The society for Research into Higher Education And Open University Press.

Ramsden, P. (1992) *Learning to teach in Higher Education*, Routledge.

Schön, D. A. (1987) *Educating the Reflective Practitioner*, Jossey-Bass, San Francisco.

Tenhula, T. (1996) *Improving Academic Teaching Practices by Using Teaching Portfolio – The Finnish Way to Do It*, ICED Conference 'Preparing University Teaching'.

Walkinton, Christensen & Kock (2001) Developing critical reflection as a part of teacher training and teaching practice, *European Journal of Engineering Education*, Vol. 26, pp 343-350.

Wiske, M. D. (ed.) (1998) *Teaching for Understanding – Linking Research with Practice*, Jossey-Bass.

Chapter 6

The Organisational Aspect of Faculty Development

Anette Kolmos, Vidar Gynnild and Torgny Roxå

1. Introduction

During the last 10 to 15 years, centres for faculty development have been established at the Scandinavian universities.

At Lund University the unit for faculty development was started in the 70'ies but it grew rapidly from the end of the 80'ies and forward. At the end of the 80'ies the unit for faculty development was established at NTNU, and finally Aalborg University followed with Centre for University Teaching and Learning in 1995 (Kolmos et al., 2001).

The establishment of the centres and their activities had the following in common:

- the decision to establish the centres was taken by the university management
- the centres were placed in the administration and were non-research-based
- the main task for the centres was to organise and convene pedagogical courses, seminars and workshops for the teachers
- activities connected to evaluation of the students
- establishing activities to improve the pedagogical qualifications of the teachers.

During the 90'ies quality development was on the national agenda for the universities. These external conditions influenced in various ways the development of the established faculty development centres, their function within the organisation, their internal organisation, faculty development strategies, main activities and their connection to research. This article will focus on the function of the faculty development centres, their organisation and strategies, and discuss the dilemmas and pitfalls which are connected to working with faculty development.

2. Development of Universities

The university as an institution goes through a violent process of change these years. Up to the 70'ies the university was characterised as a traditional academia, where research was given pride of place and the research-based education consisted in a presentation of the research results. During the 70'ies and 80'ies the university concept was widened, partly through the integration of a number of new educational programmes and partly due to the establishment of new universities, which often were given the task of developing new pedagogical models and learning methods focusing on the students. This was the case at Maastricht University, the Netherlands, Linköping University, Sweden, Roskilde and Aalborg Universities in Denmark. The research-based education changed from presentation of results to method based education, among others due to a demand for interdisciplinary and more problem-based education.

During the 90'ies the development of the universities was mostly characterized by a growing internationalisation and globalisation. In the same period the number of students increased, and the students became a more non-homogeneous group than previously, as an increasing number of grown-ups and foreigners were matriculated at the universities. University education was no longer just for the elite, but was turned into mass education, and the pedagogy should no longer be directed only towards the elite but also towards a more common level.

The global perspective also meant larger international competition, and together with national initiatives to ensure the quality of the education it initiated the development of quality procedures to improve the quality of the students' learning.

(Bowden & Marton, 1998) view this development as moving from the university of teaching and research to the university of learning. The university of learning has the responsibility of educating students who are capable of accomplishing tomorrow's tasks, and who are also capable of coping with an unknown and changing future. The university fulfils this obligation by working to a larger degree with the development of competencies.

No matter where a university is in the gap between academia and learning university, all universities have to act according to society's changed demands for their production of graduates. The knowledge society demands increased competencies and lifelong learning. The technological advances happen so fast, that within the technical fields knowledge is outdated in 5 to 7 years. So the universities both have to ensure that their own graduates obtain lifelong learning competencies; and they also have to co-operate with the surrounding society on further qualifications of the labour force.

(Barnett, 1996), however, emphasises the danger of the university's role changing from the traditional academia that builds on fundamental disciplines and the search for truth, to the new role which emphasises competencies much more, employability, lifelong learning and interdisciplinarity. This focus may be too instrumental and short-sighted, and it shifts the focus from understanding to action, where there is a risk that everything will be formulated in behavioural terms. One could certainly work with a competence approach, but it is important to maintain the critical approach to the production of knowledge and competencies as part of lifelong learning and continuous education.

Through all periods of development of the universities, research has played a dominating part. At the traditional university research is the dominant part, and often research imparting education where the teachers present their research results in the education. At the learning university research is not less important but probably it plays another part, as research forms the basis of the development of knowledge and methods, which are an essential part of the education and the students' learning.

In a research-based culture it is difficult to obtain and sustain the respect for teaching and learning, if the individual teacher is evaluated on research criteria. The system's set of values are very important to the teacher – and if teaching should be valued and respected, let alone development of pedagogical qualifications for teaching, the system has to inform about it – and render the criteria for evaluation of university pedagogy visible. And here is a dilemma, because this is not always done. The Scandinavian universities are still in the process of making the university pedagogy visible and improving its status. Lund University prepares an important way by establishing the pedagogic academy, where prestige is connected to pedagogy. If prestige, credit or necessity is not connected to the pedagogical competencies, it is a very hard task to motivate the staff to participate. It is not an easy task for the faculty units to be in charge of the pedagogical competencies of university teachers in a culture, which does not attach value to pedagogy. This is a problem which both faculty development units and the university management should be very much aware of, and which they should join forces to solve (Trowler, 1998).

3. New Management Structures

Even though the universities are in a process leading towards the learner centred university the democratisation of the learning processes is not reflected in the organisations of the universities. In Scandinavia there are very large differences in management structures. There has been a tradition for democratic elections of leaders for all posts, but changes to this practice are slowly coming about.

In Sweden leaders are still found by election, except for the rector who is appointed by the government on recommendation of an election committee at the university. There have also been changes with regard to the university board where the teachers and the students no longer are in the majority. The rest of the management is elected by teachers and students.

In Norway the "Kvalitetsreformen for høyere utdanningn" (quality reform in higher education) stated that more competencies should be given to management, and that much importance should be attached to pedagogical qualifications when new teachers are recruited. The university management has taken steps to try out different organisational experiments, but it depends on the ministry's consent. The proposal implies that the external board will get an external chairman appointed by the ministry, and that a rector, who is responsible for all activities at the institution, is appointed. All leaders will be appointed by their superiors, whereas head of departments are elected according to the present rules. The central management at NTNU will then be rector, pro-rector and university manager. At faculty level it will be the deans, the pro-deans and the faculty manager, and at department level it will be the head of department. If the proposal is approved the present elected rector, pro-rector, deans and pro-deans will be unemployed at the time when the corresponding appointed leaders take up their appointments.

In Denmark there have been democratic elections for all decision-making bodies and leading posts. But a new statute is being implemented, which will result in external boards and appointed leaders (rector, deans and head of departments). Which influence this will have on the development of university pedagogy is uncertain, but we assume that the universities will become more dependent on their managers, and that management's attitude towards university pedagogy will become crucial in the future.

At all universities there are strong indications that the management will be strengthened. (Askling & Stensaker, 2002) see it as an example of New Public Management and criticise this development. They think that a management in the traditional academic form will be absolutely capable of handling the changes that will happen in the future.

4. Faculty Development Units

Faculty development centres have been established in order to take the responsibility to train the teachers to become better teachers. Nearly all centres were established by the central university management, and most of them were part of the central administration in the beginning. This had a number of advantages and disadvantages. The advantages were that the centres were close to the central management, and the disadvantages were that the staff often saw the centres as part of the management rather than a professional unit for quality development of the educational programmes.

Co-operation with the top management is a necessity, if the institution as such is going to develop a pedagogical faculty strategy. The management seldom does it, itself, if it is not an external demand, and much literature written by faculty developers is directed to the management level (Ramsden, 1998; Brew, 1998; Centre for Higher Education Practice, 1999).

Conflicts are often seen between faculty development units and leaders at different levels, even though both parts are aiming at the same targets. For example it is not unusual that an essential reason why the university management decides to establish a faculty development unit can be found in the external publicity value just as much as in a real wish to improve faculty development. Therefore, the faculty development units have to fight for support, resources and strategic pedagogical aims – a struggle that should be fought by management – but which may end in a fight against the management. Or the unit has to fight in order to develop the quality in university education, although the responsibility is clearly placed at the top management level. Relations between faculty development units and management are therefore often very complex.

Very often, there is a lack of awareness of roles and functions both at the management level and in the faculty development units. Faculty developers take too much responsibility for developing the quality in university education, and there can be a lot of resistance to handle in the system.

5. Organizational Strategies

Even though most faculty development units have been organisationally placed in the central administration, there have been different practices as to whom they work together with in the organisation. For example the experiences from Centre for University Teaching and Learning, Aalborg University, show that even though it was the university management, which decided to establish the centre, the strategy of the centre has been directed towards the teachers rather than towards co-operation with the management. In the mid 90'ies the reason for this was among others that if the centre worked too close with the management, it would be considered the auxiliary arm of the management. So the first projects, which were launched, were based on the needs of the teachers rather than the needs of the management.

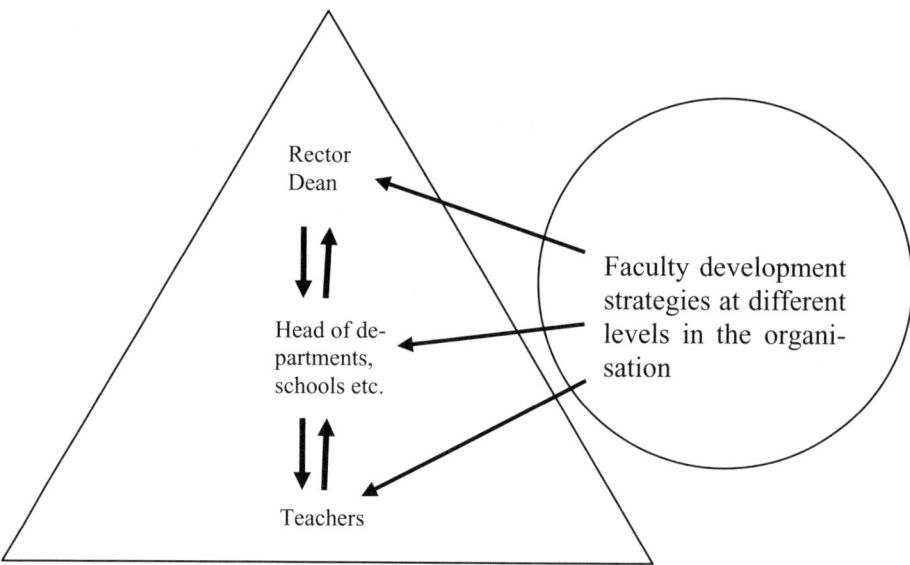

Figure 1. Relation between faculty development units and the university organisation

There is a basic difference in working as an adviser to the management compared to working as an adviser to the teachers in order to improve their teaching. The content of the job, the position in the organization and the space for action differ to a large extent – depending on the management's relation to the teachers. If there is a close relation between the teachers and the management, there may not be much of a difference, but if there is a distance between the management and the teachers there will often be some resistance towards the initiatives of the management.

Therefore, it is absolutely necessary to win the confidence of the organization right from the beginning. That is, to support projects among the teachers, to listen to the teachers and the mid-level leaders of the organization and be at their disposal with the necessary resources. Resources do not just have to be a question of money, but in particular knowledge too, like assistance in writing project applications (external as well as internal), as it is some times difficult for teachers from different subject areas to formulate problems and methods in a more pedagogical field.

The ideal situation is that faculty development units relate to all levels in the organisation - both the top, middle and bottom level, however, in order to develop that kind of practice, resources and awareness is needed. Resources enabling the units to act at all levels, e.g. to join meetings for head of departments, study boards, deans etc., and awareness of the roles of both the faculty development units and the leaders at all levels in the university organisation.

6. Research-based Faculty Development

It is characteristic for the faculty development centres established at the central administration that they are non-research-based centres. There are both advantages and disadvantages connected to this but basically there are more disadvantages. The advantages are quickly listed as it is of course cheaper for the institution if it does not have to pay for research and there will not be any competition in terms of research between the university staff and the staff at the faculty development centres. However, the disadvantages overshadow the advantages by far, as university staff will not respect staff who does not do research like themselves, and on a short view the faculty development staff will lack knowledge within their specific field and will only be able to play a superficial part.

This has become a recognised problem, and we see a tendency to move faculty development units from the administration to the department of education. This happened both at Aalborg University and at NTNU.

In the beginning"Pedagogisk avdeling" (the faculty development unit) was part of Sentraladministrasjonen (the central administration) at the then called NTH. The"Universitetspedagogiske enheten" (faculty development unit) is today named "Sektion for universitetspedagogikk" (department for faculty development) and is part of the "Program for lærerutdanning" (programme for training of teachers), which is part of the "Fakultet for samfundsvidenskab og teknologi" (faculty for social science and technology). The distance to the management of the university has grown, and physically the unit is now situated at another campus. The re-organisation took place in 1999. One of the arguments in favour of this solution was that a subject based unit should be placed together with another subject based unit. A co-ordinator has been appointed to co-ordinate the daily work in the department, which mainly consists of pedagogical qualification of newly appointed teachers, evaluations, consultations, seminars, workshops, work with action research projects etc.

At the university there have been some internal disagreements as to which are the most appropriate way of organising the faculty development unit. From different environments at NTNU very strong arguments were advanced against transferring faculty development to the teacher training programme. The department for faculty development itself proposed that a subject based, independent unit should be established, which should report to the university management. NTNU's teaching strategy makes the management's role and responsibility in development work topical, as it is seen at other universities, too. Professor Graham Gibbs at the Open University writes the following about the organisation of this activity:

"Most educational development centres are under a 'pro Vice Chancellor' (a person one level below the top who is responsible for teaching and learning matters.) Sometimes they are in a unit which provides services, which support learning, including the library and computer services. Very occasionally they are in a Personnel or Human Resource Development (this is usually a disaster). The important thing is that they are not inside a Faculty or School and that they are not associated with education." (Unpublished).

It is therefore not unproblematic to move the centres to traditional educational departments, because it means that they have to work under some very traditional research values, characterised by predominantly theoretic research without relation to practice. Exactly the combination of theory and practice is crucial in faculty development, because if the faculty developer cannot relate theory to the practice of the teachers, their prompt reaction is that the contents are useless. Many new concepts and ideas will seem frightening. Furthermore, the distance to the management will grow, and that may turn out to be a problem in the long term.

Faculty development units should be research-based and aim at establishing processes and changes. Their profile should not be basic research that is purely theoretical research, but research combining theory and practice, which has its starting point in the teachers' world. Action research is often formulated as a strategy in this field both as a qualification strategy for teachers and for staff developers.

7. Conceptual and organizational change

(Prosser & Trigwell, 1999) found links between teachers' conceptions of teaching and their students' learning strategies. Teachers who focused on information transmission tented to encourage students who used a surface approach in their studies. Teachers who instead focused on their students' learning tended to encourage a deep approach. Deep approach is associated with a strategy aiming for understanding and to integrate new information into previous knowledge. Surface approach is more or less the opposite. It is associated with memorising and focus on the examination. Deep approach is associated with better learning

In staff development (Ho, 2000) has used these findings when designing staff development specifically aiming at a conceptual change among the participating teachers. She also found, after rigorous evaluation, that teachers who did go through a conceptual development process changed their teaching strategies and consequently encouraged higher quality learning and deep approach, among their students (Ho et al., 2001). Other writers support this strategy (Entwistle & Walker, 2001), but they also describe several problems concerning teachers' sense of relevance (Trowler & Cooper, 2002).

Again these findings advocate a perspective on staff development as highly context dependent. University teachers taking part in staff development must experience relevance. If they do not feel that "their own story" is taken serious and used as a base in the process they will most likely turn away from staff development. But, on the other hand, they will engage if they are acknowledged as professionals with a serious interest for improvement and they see the staff developer as a facilitator in this process (Taylor & Bond, 1995).

It is therefore very important that faculty development has its starting point in the world of the teachers and not in an abstract theoretical world, but it is also important to relate it to strategies on organisational changes.

(Colet, 2002) operates with three different models for faculty development strategies, (Figure 2). The up-front model focuses on the improvement of the individual teacher with the development of a series of certified programmes. Typically, it involves non-research-based centres situated in the administration, and teachers who are qualified in staff development. This strategy can be applied at all levels in the university organisation.

	Up-front model	Interactive model	Distributed model
Objectives	Training of the individual teacher	Improve the system for more effective learning	Community of learning
Focus	Better teaching within the existing teaching methods	Faculty development and enhancement of teaching effectiveness	Action research projects and curriculum development projects
Activities	Training courses and comprehensive certified programmes	Counselling and curriculum development projects	R&D in the field of teaching and learning
Type of institutional centres	Non academic teacher training centres	Non academic teacher training centres	Academic centre for university teaching and learning
Organisation	Part of administration	Part of administration or research departments	Part of research departments
Type of faculty developer	Professional teacher qualified in staff development, university pedagogy and adult learning	Professional qualified in university pedagogy, faculty development, counselling and formative evaluation	Academic faculty developer focusing on creation of reflective practice
Relation to the university organisation	Minimum: Top level	Minimum: middle and bottom level	Top, middle and bottom level

Figure 2. Models of faculty development (Vinther and Kolmos 2002, developed by inspiration from (Colet, 2002))

The interactive model is a direct development of the up-front-model. Often they are used simultaneously at the institutions. The interactive model focuses on development of the system, and therefore it involves both counselling and curriculum development projects. Typically, it involves non-research-based centres located at the department level, and it usually leads to a higher qualification level for staff and faculty developers compared to the up-front model. Compared to the up-front model, the interactive model builds more on the interaction and the dialogue between staff / faculty developers and the ordinary teachers. In other words, there must be two actors involved to make this strategy function, and at first it is necessary to establish a discussion and a pedagogical reflective practice (Lave & Wenger, 1991; Wenger, 1998).

The third model is the distributed model which focuses on creating a new system and culture referred to as a community of learning. This model is based upon research-based faculty development in order to facilitate action research projects run by faculty developers and staff, and it involves counselling regarding curriculum projects. Research-based faculty development is a requirement so that the staff and the faculty developers do not only end up in pure pedagogical and educational pragmatism, but rather in evidence based development. This model requires a high level of integrity among all actors in the organisation and the faculty development unit - and requires an experimenting culture and an action research approach combining theory and practice. This is a very important aspect, as many faculty development units have been merged with faculty of education. The marriage has not always been a happy one, as it may end up in a clash of two research cultures - the traditional educational research and an action oriented and applied teaching in university pedagogy.

Much development in higher education is general and is based on trial and error, in spite of the paradox that it often occurs within research-based cultures. Action research is an obvious method for the improvement of students' learning, because experimentation can coincide with evaluation, which paves the way to the creation of a reflective culture. Learning is such a complex matter and is not becoming any less complex in terms of achieving new competencies for lifelong learning.

8. Three Cases

The following three cases are in various ways illustrative of the background for the function and development of the faculty development units.

8.1 The Lund case

In **Sweden** higher education has been decentralised during the last decade giving the various institutions the possibility to build individual profiles. This has been followed by an increasing external evaluation securing accountability and quality. At Lund university this process has been mirrored in the university organisation where the faculties have developed rather individually in relation to staff development and other teaching development initiatives.

One of the faculties, Lund institute of technology (LTH), has reached for this possibility and turned teaching and learning into one of its core profiles formulated in the Breakthrough project. Below the historical background for the project will be described and discussed.

In the beginning of the 90'ies the then recently published national inquiry into Swedish higher education proposed pedagogical training for all university and college teachers. The government did not approve of this, but considerable funds were provided for the enhancement of pedagogical competence among teachers. At LTH this lead to the establishment of various "pedagogical courses", which were offered to teachers on an individual basis.

The courses were "bought" from the centre for teaching and learning at Lund University. This customer and provider relationship between the centre and the faculty was both unique and important when it comes to understanding the development that was to follow. Because of this buying and selling, the cost for courses and consultation showed in the faculty's budget every year. This provoked the Dean and other managers to discuss the value of the activities every year. The provider, the centre, on its part had to develop sensitivity towards the faculty and its specific needs. Therefore, the courses focused more and more on the needs perceived by teachers. On the other hand, the Dean and the head of departments - as expressed by participating teachers - noted the value. Because of this, the courses earned quite a high status and interest in the organisation in the long run.

It was followed by an increase in the demand for consultative support. Teachers who had taken part in the courses initiated seminars, curriculum development, and other initiatives in their departments and asked for the participation of the particular consultant whom they encountered during the course. The demands on this service increased. It also involved demands from head of departments, study directors and the Dean on services concerning enhancement of quality in teaching and learning. Several of these initiatives were realised in different parts of the faculty, both development for individuals and development of entire study programmes.

The process was then formulated in the Breakthrough project, a large-scale development initiative within the faculty. The objective was to increase collaboration between different parts of the faculty, mainly by creating a teaching and learning culture based on a learning centred approach. This was to be achieved by critical and informed conversations between teachers where they investigated their own practice. The fact that the teachers' practice was in focus must be considered as important. The project was to succeed if the teachers were motivated to participate in the building of a culture. They were not to implement anything, nor were they only asked to learn about learning. They were about to develop their own practice by forming a culture together with colleagues.

In this the staff development approach is very much like that of a facilitator. Teachers, head of departments and the Dean are encouraged to share their experiences about their own practice. The developer's focus is to support this story telling, to challenge conceptions and ways of understanding, to scaffold any attempts, on the teacher's behalf, to formulate development, and to direct attention to literature, to examples to follow, to colleagues to talk to, etc. The teacher, the team of teachers, the Dean, or whoever, should experience a sense of ownership of the process. And this is important. It is the practitioner's practice that has to be developed. It will never ever develop further than the practitioner allows it to develop.

The outcome of this is described in a separate chapter in this book. Here, we will only mention an example:

- About a quarter to a third of all teaching staff has been actively engaged in staff development activities during the last twelve months. And many more in informal discussions within their departments.

- A focus group consisting of the head of departments indicated an increased discussion among teachers. This both in quantity and quality, both within subjects and across subject boarders.

- The head of departments also indicated that the curriculum was changed in such a way that the students worked with applied subjects earlier in the programmes, and that the use of projects as teaching methods increased.

- Individual teachers indicated, in replies to questionnaires, a development of a language about teaching and learning. They also used this language while reflecting at their teaching practice and while talking to other teachers. However, they also indicated that they did not use the vocabulary in all conversations, but only when discussing with colleagues who had taken part in staff development.

- LTH's pursuit of forming a leading role in the development of teaching in Swedish higher education has, among teachers, begun to take the form of an objective possible to accomplish.

But there is still work to be done. The principles described above will continue to guide the process. The lesson learned earlier in the 90'ies, when staff development services were sold to the faculties, resulted in two insights. Firstly, the importance of displaying the cost is vital. Staff development costs money and time in terms of teachers' working hours. It is when the Dean and the head of departments become aware of this, that they also take a position and formulate the purpose of staff development. Secondly, the consumer and provider relation between the centre for teaching and learning and the faculty and its teachers helped focusing staff development activities on the teachers' practice. That was the only way of keeping them as costumers.

This way of reasoning might be a bit odd in a higher education context. On the other hand, quality in these matters should not be confused with only satisfying the client's needs as understood by him or her. Working in a scientific context means that everything is scrutinised and searched for value. The only way of keeping scientists as customers is to go beyond their own expectations and help them to discover things they did not know about. That is: Staff development in higher education must be intellectually intriguing and challenging and still be close to the practice it is meant to develop.

8.2 The NTNU case

The faculty development work at NTNU in Trondheim is strongly influenced by the lines written in the so-called "Kvalitetsreformen for høyere utdanningn" (quality reform in higher education) (Stortingsmelding 27 (2000-2001 (*Gjør din plikt – krev din rett)*). The primary objective of the reform is better quality in higher education. Among others the institutions are going to develop a system for quality control (Giertz, 2000). Several persons from the Department of faculty development take part in subcommittees working at implementation of the quality reform at NTNU. In addition to this the department takes care of a systematic training of newly appointed teachers, but larger importance is going to be attached to service and innovation in the subject-based environments. The department is also involved in supporting the implementation of new teaching and evaluation methods, and the training of assistant professors.

It is still discussed at NTNU if the Department of faculty development should be closer related to the university management. At present, the department is placed within a faculty, and among others this has proved to be rather inconvenient from a resource point of view. The co-ordinator at Department of faculty development finds that it is now time to improve services within two – three main areas. Among others there is a wish for increased competencies within subject didactics and ICT-didactics, especially in order to meet demands in the engineering educational programme. Based on the work with the quality reform, proposals have been made for the establishment of a pedagogical information service combining more subjects than the present Department of faculty development.

When it comes to practical tasks the department is among others responsible for assistance in evaluation of teaching and learning. The evaluation has been a permanent arrangement at the civil engineer educational programme for many years. Special attention has been paid to oral, formative evaluation in the so-called reference groups, which consist of a few students and teachers from each subject. Systematic, scheduled final evaluations have also been carried out. Due to the quality reform the evaluation system is being revised.

Another central task is to carry out a pedagogical development programme for newly appointed teachers. It is compulsory for the permanent staff and it consists of two parts: A common part (64 hours) and optional modules (36 hours). The total number of working hours is calculated to about 100 hours. Normally, the programme is divided into two semesters in addition to the teachers normal job, but the optional part of the programme can be divided into four semesters without special application. The common part of the programme consists of gatherings, a lecture in colleague based supervision and a development project connected to the teacher's own teaching situation parallel to the programme. The project is finished with an individual report.

The common part of the development programme is intended as a contribution to make the university teacher develop his/her own pedagogical competencies. It implies among others that the teacher acquires theoretical knowledge as the basis for analysing and solving pedagogical challenges in connection with his/her own teaching. On admission to the development programme teachers belonging to the permanent staff at NTNU is given priority according to their seniority.

The optional part of the pedagogical development programme consists of large or/and small modules at free choice, which in all extend 36 working hours. To meet the demands the participants should have either a large and a small module or three small modules. Large modules extend 24 working hours and small modules 12 working hours. The modules mostly consist of two or three gatherings. New ideas and methods are tested in the teacher's own teaching between the gatherings. The participants report on this at the last gathering. At present the following optional modules are offered:

Large modules:
- Problem-based Learning (PBL)

- Basic course in PBL-supervision
- Examination and alternative ways of evaluation
- Teaching large groups
- Students' evaluation of teaching and learning

Small modules:
- Subject-based supervision
- Problem-based Learning (PBL)
- ICT in learning

8.3 The Aalborg case

In Denmark there were especially two ministerial regulations which had great influence on the establishment of the pedagogical units. 1) During the period a ministerial evaluation centre was established the task of which was to make evaluations by rotation, in order to compare and evaluate university education at various national universities. This meant that more focus was attached to the internal quality development at the universities, and at the same time it motivated the establishment of curriculum development projects. 2) A new law concerning position structure was implemented, which meant among others that assistant professors were to be supervised and guided as to their teaching. The law was implemented with a decentralised strategy as each university could develop its own programmes to meet the legal demands. This resulted in a series of pedagogical courses for assistant professors at the universities, but without common national certificate only certificates issued by the various institutes.

External and internal quality assurance
Centre for University Teaching and Learning was established in 1995 on the basis of a decision made in the Academic Council. At the time there was widespread scepticism in the organisation towards this initiative, and therefore the centre chose to work with a bottom-up strategy rather than with a top-down strategy right from the start. However, this meant that communication between the centre and the management could have been much better than it was in the first years. Thus the centre did not regularly participate in central meetings for deans, head of departments, study boards etc. In the beginning a round of visits was arranged to collect the needs of the institution but no systematic follow-up was made. This was among other things due to a lack of resources, but in the last resort it was rather a question of deciding priority of the available resources in the faculty development unit.

Due to the introduction of external evaluations there has been ample opportunity to involve the Centre for University Teaching and Learning in the work with internal evaluations of the study programmes or in the reverse to let the centre define its own role. But at the institution the case has been a parallel course of events where study boards have made their own evaluations and only to a limited extent made use of the pedagogical expertise. There have not been any conflicts between the various management levels in the organisation and the centre – on the contrary. But the awareness of a strategic approach in relation to the university organisation has come around after some years, and now it is obvious that pedagogical development has to be accomplished at all levels in order to establish an interactive and distributed model.

Compulsory training
One of the largest activities is the compulsory pedagogical education for assistant professors, which extends over 175 hours for each assistant professor. Since the beginning more than 300 assistant professors have completed the course. The organisation of the education has been a contributory cause to the success in bringing faculty development up for discussion in the organisation - and especially in preparing the ground for the establishment of curriculum development projects and subject based pedagogical courses. So the compulsory education is an essential part of the strategy for development towards an interactive and distributed model.

The goal for this course is for the assistant professor to gain knowledge about the theoretical educational and didactic methods, leading to reflection while teaching, and about the development of the learning processes. Furthermore, the assistant professor will gain the ability to develop, plan, implement, and evaluate different types of educational programmes while promoting the strengths of the learning process. The educational foundation of the course combines theoretical reflection with concrete teaching methods. Knowledge and development of professional didactics comes to play a central role in the course. Thus the intention of the course is to provide an environment conducive to participant-directed and experience-based learning processes for the individual assistant professor. The course extends over 3 semesters. There are 4 combination modules where there is an exchange between theory and practical application. The assistant professor's contribution is estimated to be 175 working hours.

Each assistant professor is allotted 2 advisers: one adviser from The Centre for University Teaching and Learning and a practical adviser assigned by the assistant professor's own department, formally appointed by the head of the department. This model has had the effect that not only more than 300 assistant professors have attended the course but also 300 advisers (associate professors and professors) have attended the courses. Generally, it is almost impossible to make associate professors and professors attend activities, but if they are involved as experts they like to participate. And the motivation for learning about university pedagogy is very large because it is difficult to supervise the assistant professors without some understanding of the fundamental pedagogical / didactic concepts and without reflecting on one's own teaching experiences. Therefore, both the assistant professors and their advisers are offered a number of workshops in the field of university pedagogy.

It is compulsory training – and the effect is beginning to show in the system. Many young associate professors are seriously beginning to experiment with pedagogy, to make extensive curriculum development projects etc. To participate in the course gives them a basic pedagogical qualification and an idea of enhancement of learning and teaching.

9. Perspectives

As can be seen from the above three cases, there are a number of dilemmas to which faculty development centres at universities in Scandinavia will have to find solutions for in the years to come.

It is a general characteristic in the three countries involved, that they have **national decentralised strategies** for faculty development. There is no common national certification level, as you see in other countries like Baden Württemberg, Germany, where it is the ministry which has launched a certification process and development of national courses. In the Scandinavian countries there is a tradition for democracy and influence, so the national control and regulation uses other means to put a pressure on the pedagogical development on institute level. The most important advantage by the decentralised strategies is that they involve the individual organisation to a larger extent, and that the management and the staff are made responsible for making decisions about their policy in the field. This strategy also makes it possible to practice the interactive and distributed model of faculty development, where the point is to make pedagogy a natural element of a usually research- based culture.

On the other hand, the national decentralised strategies also imply disadvantages like large differences in the certification processes of the various institutions as to both content and level. This can be handled to some extent within the country's own boarders as the variations in culture are more or less well-known. But internationally it will cause problems in the long run, because there will be too many and too specific certification processes.

On **institution level** there is a clear tendency that universities are changing management structures from elected leaders to appointed leaders. It is difficult to predict how this will affect faculty development in the future, but we may fear that university systems with a more hierarchic management structure will be difficult to change, which will influence the liberty of action for the faculty development units.

There are some common characteristics as to how faculty development centres have been started, and the organisational considerations of how the centres are to interplay with the whole organisation. The contact to the management is a prior condition to bringing pedagogy up for discussion in relation to strategies for the universities. But it is also necessary to work among the teachers to motivate each teacher to work with improvement of quality in education. Therefore it is a both / and situation which demands resources and deliberate planning.

At present, there is a tendency to remove faculty development centres from the management and place them at institutions for education. It is a very logical and rational solution which may contribute to solve the other large problem of the centres, which is their lack of research. But it creates two essential problems: firstly, that the centres are removed from the management and that the system sees them as part of an institute for education, no matter which relation the actual institute may have to the rest of the organisation. Secondly, that the research carried out by these institutes is more theoretical than the research needed to develop university pedagogy. University pedagogy should be research-based but with focus on the combination of theory and practice.

Research-based university pedagogy will probably solve the problem that the faculty development units seldom document the processes and the results of their efforts. Thus the practice of changing education programmes without real documentation is supported. There is no doubt that all three universities described in this article see the aim of faculty development within the frames of an interactive and distributed model. The very strategies based on the teachers' practice and consultant work underline this. But development towards a university culture, where faculty development becomes a natural element, demand action oriented research methods and documentation of the processes of change, where not only faculty development units do research in the processes of change but where the teachers do research in their own teaching processes to a larger extent. There are elements of this in the Scandinavian faculty development culture, which may form the basis of further development.

References

Askling, B. & Stensaker, B. (2002) Academic Leadership: Prescriptions, practices and paradoxes, *Tertiary Education and Management* Vol. 8, pp 113-125.

Barnett, R. (1996) *The Limits of Competence - Knowledge, Higher Education and Society*, The Society for Research into Higher Education and Open University Press.

Bowden, J. & Marton, F. (1998) *The University of Learning*, Kogan Page, Göteborg.

Brew, A. (1998) *Directions in Staff Development,* SRHE and Open University Press, Buckingham, pp 21-35.

Centre for Higher Education Practice, Open University (1999) *Institutional Learning and teaching strategies: A guide to good practice*, HELFE, Bristol.

Colet, N. R. (2002) *Faculty development strategies,* Paper written for the STRATA ETAN Expert Group, fortcoming.

Entwistle, N. & P. Walker (2000) Strategic alertness and expanded awareness within sophisticated conceptions of teaching, *Instructional Science,* Vol. 28, pp 335 – 361.

Ho, A. S. P. (2000) A conceptual change approach to staff development: A model for programme design, *International Journal of Academic Development*, Vol. 5, No 1, pp 30-41.

Ho, A. S. P., Watkins, D. et al. (2001) The conceptual change approach to improving teaching and learning: An evaluation of a Hong Kong staff development programme, *Higher Education*, Vol. 42, pp 143-169.

Kolmos, A., Rump C., Ingemarsson, I, Laloux, A and Winther O. (2001) Organization of Staff Development - Strategies and Experiences, *European Journal of Engineering Education*, Vol. 26, No 4, pp 329 – 342.

Lave, J. & Wenger, E. (1991) *Situated Learning*, Cambridge University Press.

Prosser, M. & Trigwell, K. (1999) *Understanding Learning and Teaching. The experience in Higher Education*, The Society for Research into Higher Education & Open University Press.

Ramsden, P. (1998) *Learning to lead in higher education*, Routhledge, London and NY.

Taylor, P., Bond, C. et al. (1995) *Academics as learners: issues for courses in the study of higher education*, Improving Student Learning: Using Research to Improve Student Learning, Exeter, The Oxford Centre for Staff Development.

The Ministerial Circular on Job Structure for Academic Staff working with Research and Teaching at Universities, etc. under the Ministry of Research and Information Technology (2000).

Trowler, P. & Cooper, A. (2002) Teaching and Learning Regimes: Implicit theories and recurrent practice in the enhancement of teaching and learning through educational development programmes, *Higher Education Research & Development*, Vol. 21, No3, pp 221-240.

Trowler, P. R. (1998) *Academics Responding to Change in New Higher Education Frameworks and Academic Cultures*, Society for Research into Higher Education and Open University Press.

Vinther, O. &Kolmos, A. (2002) National strategies for staff and faculty development in engineering education, *Global Journal of Engineering Education*, Vol 6. No. 2.

Wenger, E. (1998) *Communities of Practice - Learning, Meaning and Identity*, University Press, Cambridge.

Chapter 7

Proposing Nordic Excellent Teaching Practice, NETP

Pernille Andersson and Lauri Malmi

1. Abstract

This paper presents the idea of developing the Nordic Excellent Teaching Practitioner, as a tool for advancing and evaluating engineering educators' teaching skills in the Nordic countries. NETP would be based on the preparation of teaching portfolios, which would be evaluated using common criteria and similar processes in Nordic technical universities and institutes of engineering education.

2. Introduction and vision for the Nordic ETP

Currently, higher education keeps ahead of the challenges of how to educate more students with more heterogeneous backgrounds with fewer resources than before. In addition, the content of many subjects becomes more and more complex. Moreover, the working life requires higher demands from the students. They should be able to handle complex situations, teamwork and be self-managed from the very beginning of their career.

These facts naturally require new demands from the persons, i.e. teachers and study officers, responsible for the organisation and implementation of higher education. Thus there is a great continuous need for staff and faculty development and for encouraging teachers in higher education to improve their skills in teaching and learning. A number of successful projects and processes concentrating on this issue has emerged in technical education in

Nordic countries, for example, the YOOP-education in Helsinki University of Technology (15 credits) (Teaching and Learning Development at HUT, 2004.), and the Breakthrough project and Pedagogical Academy at Lund Institute of Technology at Lund University (Hammar Andersson et al., 2002). The last-mentioned project is described in articles in this publication. Experience gained in such processes should not remain only in the original institutions. Instead, it ought to catalyze new activities elsewhere in order to find and provide opportunities for teachers to learn more about teaching and student learning.

Many challenges teachers meet in higher education are universal, that is, teachers in universities and other higher education institutions in different countries face similar problems. One of the best solutions to meet these challenges is to share experiences and best practices in teaching and learning, because it stimulates and develops new ways of thinking and new ways of handling the challenges. Co-operation between colleagues in the same department or at a whole university is both relevant and beneficial for the development of higher education. Moreover, co-operation between universities at a national and international level is desirable in order to disseminate successful experiences and practices even wider. As an example, consider the IPN network in Denmark (IPN, 2004) or the TIEVIE network in Finland (TIEVIE, 2004). Incidentally, it is surprising that teaching in higher education is often considered a personal task (where such co-operative methods are not needed), even though discussion and critique on new ideas and results is more natural working method in research. However, we should never forget that learning always is a social activity and the sharing of thoughts is a main road of the academic profession.

Current trends in higher education promote competition between research fields, institutions and countries. Thus developing the competences of staff and faculty members should be a major issue for universities and the whole academic society. It is therefore obvious that the increased co-operation among the faculties and universities of technology in the Nordic countries is worth working for in many fields. In our point of view, it is important to share the efforts into developing and evaluating the competences in teaching and learning to achieve progress in the process.

This paper presents a proposal to introduce a system of a Nordic Excellent Teaching Practitioner (NETP) degree for teachers. The main purposes of the system are the following. First, NETP should stimulate teachers to develop their teaching and learning activities in co-operation with others. Second, it should stimulate the sharing of best practices and new solutions among the Nordic technical education institutions. Third, NETP should promote qualified engineering teaching with the long-term aim of assuring that all teachers in the Nordic countries have the same teaching standard.

A system like this is needed in order to motivate the teachers to improve their teaching skills and give them new challenges in their professional development. The system will also be motivating for the management level because it is beneficial to have co-operation even in this field and, moreover, to be able to exchange faculty personnel with an assurance of quality.

The main idea described in this paper is that teachers in Nordic engineering education participate in workshops with colleagues from other Nordic countries. They prepare teaching portfolios and the evaluation of the teaching portfolios will be based on criteria that mirror the demands made on quality teaching. After this process the teacher gets the competence level Nordic Excellent Teaching Practitioner. We hope that such a competence will gradually gain general appreciation when applying for new positions in academic institutions. It can also form part of the basis for decisions on promotions and bonuses to staff and faculty members.

We begin with arguing for why NETP should be based on teaching portfolios, and recall some experiences of using portfolios in evaluating teachers' skills. Thereafter we briefly discuss various aspects, which are relevant when launching such an initiative. Finally, we suggest how the process of defining NETP could be started.

3. Teaching portfolios

One technique that becomes more and more common in stimulating the development of teachers is teaching portfolios (Seldin, 1997). Teaching portfolios have many purposes. First, a portfolio is a document in which a person gathers samples of his/her teaching practice. Second, often a teaching portfolio also includes the teacher's written commentary that provide context and shows the teachers reflections over his/her personal ideas and theories of teaching and learning. The documents included in the portfolio support this by providing the evidence of how the ideas are used in teaching situations and they are the data on which the reflection is based. Finally, a portfolio also shows the professional development and future possibilities.

A portfolio can be used both as a personal and an official document. Portfolios are used for personnel decisions, such as recruitments and promotions as a document showing both how extensive teaching experience a teacher has, and how he/she thinks and reflects upon issues of teaching and learning. The portfolio is also a personal document in which the teacher can document teaching experience and monitor his/her personal development (Apelgren & Gietrz, 2001).

In the process of creating a teaching portfolio, teachers become more aware of their practice and its impact on student learning. This means that teaching portfolios can be used as tools for sharing best practices and stimulating the co-operation and dialog among the teachers. With this document they can better describe their practice in teaching and learning to each other.

The most essential idea with preparing and updating teaching portfolios is to make them a start for a continuous development of experience and knowledge in teaching and learning. It is about gaining new knowledge form reflecting upon one's practice. This is one of the main thoughts in the scholarship of teaching (Boyer, 1990).

3.1. Developing knowledge

An important issue in the process of writing a teaching portfolio is that teachers train themselves in describing and giving arguments about their educational activities with adequate pedagogical terms. This is an opportunity to rethink the ideas of teaching and learning. Going into this process also gives opportunities to learning from colleagues' experiences and thinking.

During the construction of a portfolio teachers can gradually get a clearer picture of how their teaching has evolved over time, the context they work in and its impact on their actions and thinking. They also become more familiar with their own strengths and weaknesses.

Teaching skills are often what is called tacit knowledge developed by trial and error and never made really conscious (Schön, 1984). They often also depend on the context in which the teaching activities take place (Lindén, 1998). Writing a teaching portfolio starts a process of making the teaching skills conscious thus turning the experience into professional knowledge not dependent of the context anymore.

This process is not always easy for the teacher to go through. Some kind of support such as workshops or courses is often necessary to assist the teachers in the process. The possibilities for co-operation and discussions between colleagues, which these activities offer, greatly support the writing process. Again, we must recall that learning is a social activity and not only an individual activity.

4. Experience from the Nordic countries

The idea of a Nordic ETP system has its origin in the system of rewarding scholarly teaching at Lund Institute of Technology at Lund University. In this institute teachers write their teaching portfolios and the evaluation is based on six criteria, such as "the personal philosophy of the applicant constitutes an integrated whole, in which different aspects of teaching are described in a way that the driving force of the applicant is apparent", "a clear development over time is apparent. The applicant should, preferably, consciously and systematically have striven to develop personally and in pedagogical activities" and "the applicant has co-operated with other lecturers in an effort to develop his or her teaching skills" (Hammar Andersson et al., 2002). The formulation of the criteria is based on the ideas of scholarship of teaching. If the teacher's teaching portfolio is approved, he/she gets the competence degree ETP and increase of his/her salary. Moreover, the department where the teacher is employed also receives a rise in its funding. Thus, the surrounding organisation also benefits directly from the increased competence of its members.

Teaching portfolios are, of course, used in many other Nordic institutes, as well. Uppsala University in Sweden has used teaching portfolios for several years as a tool in the process of recruiting and promoting teachers (Apelgren & Geirtz, 2001). In Finland the University of Oulu has been in the lead of using portfolios in assessing teaching merits. At the moment portfolios are officially used in most Finnish universities. However, the process of evaluating them properly is still progressing. In many cases, for example, in Helsinki University of Technology, portfolios mostly list only hard facts about teaching experience, and do not include a vision of the teacher's scholarship of teaching with work samples to document this vision.

In general, portfolios are gaining wider acceptance, and the evaluation of their contents will gradually improve. We believe that now it could be time also to use teaching portfolios at a Nordic level to demonstrate the competence of engineering teachers to the rest of Europe.

5. Nordic Excellent Teaching Practice

We would like to suggest a Nordic Excellent Teaching Practice to motivate teachers and management in higher engineering education in the Nordic countries to improve their competence in teaching and learning. It will also be an opportunity for teachers in the Nordic countries to meet and share experiences and form networks.

Our vision is to develop a common Nordic standard to ensure qualified engineering teaching and to assure that all teachers in the Nordic countries have the same teaching standards. We recognize that this is not an easy task and the achievement of such a goal will take place in a distant future. Nevertheless, the process should start now, later it can be refined by collecting experiences from institutes from all the Nordic countries.

The process of creating an NETP system has to address several important issues.

- What is the process of evaluating teachers' competences? It is obvious that the evaluation is a local process at each institute and therefore has to conform to local traditions and administrative processes. However, it is possible to make an overall guideline on the necessary steps.
- How teachers' competences are evaluated, i.e. what are the evaluation criteria? This is the real challenge: how can criteria that are widely accepted and at the same time not too broad and general in nature be developed. Since portfolios have been used and evaluated in many universities in Nordic countries for years, experience of their evaluation processes should be gathered and the essential aspects should be found.
- How should the process of preparing a teaching portfolio to get the NETP be supported? This includes questions such as how staff education is carried out and what contents does it have. The whole system is insufficient if we cannot provide a roadmap that shows the teachers how they can improve their competences. Again, it is not possible to set up a list of courses to be held in all institutes, but instead a list of recommended topics to be covered in staff and faculty development can be made. Moreover, a proposal of how these activities could be promoted by co-operation on both national and international level ought to be developed.
- What kind of accreditation system should be formulated to form similar types of evaluation processes at the different institutions? Questions such as how to cope with national or international level co-operation, or should we proceed towards setting up some accreditation organisation have to be addressed. Moreover, should the process lead to changes in the formal requirements of teaching positions in the institutes, and what is a feasible process of developing the accreditation activities.
- What kind of guidelines could be given to implement the system in Nordic universities and how to make it a part of normal work? What kind of national level and Nordic level coordination and co-operation is needed to promote the implementation process?

- Finally, we also face the question of what to do with the teachers that have got the NETP. One suggestion is to arrange seminars, courses, workshops and even conferences for them to keep the networks together and to continuously give the teachers opportunities to develop their competences.

6. Starting the process

This paper is a proposal to start the process of defining and implementing NETP. Our vision is that such work, if successful, would gradually promote the appreciation of teaching skills both at an individual and an organisational level. When we ask institutes to send a representative to discussions on an initiative and afterwards ask for their opinion on the NETP draft, we also declare that promoting education and teaching skills is an important international activity, which they should not ignore.

This paper originates from the first IPN seminar on staff and faculty development within engineering education in Scandinavia, held at Aalborg University in 2002. The second Nordic symposiums on staff and faculty development in engineering education will take place in Odense, Denmark, 2004. We suggest that they should work on this proposal and launch the process of defining and finally implementing NETP.

References

Apelgren K., & Giertz, B. (2001) *Teaching Portfolios*, Uppsala University, Uppsala.

Boyer, E. L. (1990) Scholarship Reconsidered. Priorities of the Professoriate, The Carnegie Foundation, New Jersey.

Hammar Andersson P., Olsson T., Almqvist M. et al. (2002) *The Pedagogical Academy – a way to encourage and reward scholarly teaching*, Oxford, Proceedings 10:th conference Improving Student Learning.

IPN, (2004) IPN - *Pedagogical Network for Engineering Education in Denmark*, Retrieved July 7, 2004 from http://www.ipn.dk/.

Lindén J. (1998) *Handledning av doktorander*, Nya Doxa, Nora.

Teaching and Learning Development at HUT (2004) Retrieved July 7, 2004 from http://www.dipoli.hut.fi/ok/yhteystiedot/documents/teaching_and-learning_development.pdf.

Seldin P. (1997) *The Teaching Portfolio – A Practical Guide to Improved Performance and Promotion/Tenure Decisions*, Anker Publishing Company, Inc., Boston.

Schön, A. D. (1983) *The Reflective Practitioner*, Basic Books, Inc.,San Francisco.

TIEVIE (2004) *Tievie – Tieto- ja viestintätekniikan opetuskäytön virtuaaliyliopistohanke* (Finnish virtual university project to support educational use of information and communication technology), Retrieved July 7, 2004 from http://tievie.oulu.fi/.

Chapter 8

Faculty Development in Engineering Education in Finland

Johanna Naukkarinen and Lauri Malmi

1. Abstract

In this paper we present the current engineering education system in Finland and how pedagogical training for teachers of engineering is organized. We mainly consider the university sector giving Master of Science in engineering degrees. However, the outline of the polytechnic sector giving Engineering (polytechnic) degrees is also described. However, we do not consider the organisation of postgraduate studies in either sector.

In the polytechnic sector there is a long tradition of pedagogical training and proficiency requirements for teachers, because the predecessors of polytechnics, the technical institutes, were mainly teaching-orientated whereas universities have always been research-orientated. However, also the university sector has placed considerable emphasis on the teaching abilities of its faculty members during the last ten years. Thus, portfolios or equivalent documents of teaching merits are required when applying for teaching positions, specially tailored pedagogical training is available in many universities, and many kinds of evaluations are carried out. There are also a number of national pedagogical activities and networks to support teaching development in universities.

2. Engineering education in Finland

The history of technical education in Finland is roughly 150 years long (Orelma, 1996). Despite of the shortness of this period technology has become the largest single educational field in the Finnish tertiary-level education (Ministry of Education, 2003a & b) measured by the number of students in the field.

2.1. Background

The first forms of formal technical education in Finland were the Sunday- and evening-schools established in 1842, whose objective was to give the journeymen all-round education to complement their apprenticeship training in professional skills. Sunday-schools were soon accompanied by three full-time technical schools in Helsinki, Turku and Vaasa. The Helsinki Technical School started evolving towards higher education objectives, whereas the others kept a practical focus of educating machinists and foremen for industrial production plants. In 1908 the Helsinki Technical School became Helsinki University of Technology (HUT), the first technical university in Finland. (Orelma, 1996). Later in the 20th century two more technical universities (Tampere, Lappeenranta) and three technical faculties (Oulu, Vaasa, Åbo)(The faculty of chemical technology in Åbo Akademi gives education in Swedish.) were established.

The more practical technical schools formed the roots of the technical institutions, which during the 1990s were changed into polytechnics and thus joined the technical universities in the higher-level educational sector. As most of the current polytechnics are multi-disciplinary, in practice this meant that the technical institutes were joined with other institutes in the same region to form a polytechnic. The last trace of secondary-level technical education, the technician degree, was abolished by Ministry of Education by the beginning of 1999 (Ministry of Education, 2000, 2).

Nowadays' technology education, great majority of which is engineering education, is the largest educational field in tertiary level. In 1999 approximately 23% of all tertiary level degrees came from the field of technology. This ranks Finland second among the OECD countries, with only South Korea educating more engineers in proportion to all graduates (Allt, 2003). Table 1 presents more statistical information of the tertiary-level technology education in Finland.

Table 1: Statistical information of tertiary level education in the field of technical sciences in universities and engineering and transportation in polytechnics (Ministry of Education, 2003 a & b).

	University sector	Polytechnic sector
Number of new students 2002	4 428	11 177 *
Number of all students 2002	36 443	42 208
Women of all students (%)	~20 % **	16,3 %
Employment of year 2000 graduates at the end of 2001 (%)	91,1 %	89,3 %
Number of teaching staff 2002 (men-year) ***	1 250	1 672
Student / teacher –ratio	29	25

* Number includes students starting in normal programmes (9 604) and in adult education programmes (1 573).
** The exact percentage of women in the university sector was not available in KOTA-database report. The number is an estimated average of the share of women starting university sector studies in 1991-2000 (Allt, 2003).
*** Polytechnic number consists of full-time teachers (1 595) and part-time teachers (adds up to 77 man-year). The university number consists of a sum of man-years of all kinds of teaching posts (full-time and part-time).

2.2. Structure and content of engineering studies

The Finnish engineering education system resembled for a long time the German-type dualistic model, where technical institutes provided the shorter and more working life orientated degrees, and universities offered the longer and more theoretically-orientated degrees. This division of responsibilities was generally accepted and both degree types had well-established positions in the labour market. The basic idea of the dual model remained even when the polytechnic sector was formed and the technical institutes disappeared (Ministry of Education, 2002a).

All university level engineering degrees consist of courses altogether worth of 180 study credits (In Finnish terminology the term "course" refers to the smallest unit the student can include into his/her degree. Typically the size of courses range from 2 to 8 study credits. One study credit corresponds approximately to 1,5 ECTS units.), where one study credit equates to about forty hours of work. Universities are required to arrange the education in such a way that it enables completion of the engineering degree in five years. In practice, however, between 1999-2001 the average study time varied between six and seven years depending on the university and year (Ministry of Education 2003a). One likely reason for the long study time is that most

Finnish engineering students work part time during their studies. The graduates from spring 2001, for example, already had an average of 20 months of work experience by the time of graduation (Allt, 2003).
Education is planned and executed as degree programmes, and students are usually admitted to the programmes, with some exceptions. In the spring of 2003 technical universities and faculties offered altogether 60 degree programmes or options, out of which four were in architecture and the remaining 56 in engineering (Teknillisten tieteiden yhteisvalintaopas, 2003). The university level education in engineering emphasises the theoretical abilities and general skills in the chosen field of engineering.

University degree in engineering is composed of basic studies, studies in the chosen field of engineering, advanced special studies in the chosen topic in the field (major) and practical training. Studies in some other field of technology are generally included in the degree, too (minor). Basic studies usually include natural sciences, computer science, language and communication studies and other general knowledge, such as law or economics. Advanced special studies also include writing a thesis (20 study credits), which most students do for industrial companies. The composition and extent of the building blocks of the degree change from one university to another. An intermediate degree in university level engineering education will be introduced in the autumn of 2005 as the Finnish educational system is steered to a more EU-complying direction (Ministry of Education, 2002b).

In polytechnics the degrees offered in the field of engineering and transportation vary from 140 to 180 study credits in extent, with most engineering degrees at 160 study credits (National Board of Education, 2003). The computational duration of engineering studies at polytechnic level is four years. The actual average time to complete the studies in engineering in polytechnic ranged between 4,2 to 4,4 years during 1997-2002. Also the polytechnic level engineering studies are organised as degree programmes. At the moment there are 31 polytechnics in Finland out of which 24 offer altogether 39 degree programmes in engineering (Ministry of Education, 2003b).

The polytechnic degree consists of basic studies, professional studies, professional training and demonstration of scholastic ability. Generally the polytechnic level degrees aim at familiarising students with the practical knowledge and skills needed in expert tasks in professional life and enhances the application of this knowledge in solving the professional problems. When compared with the more theoretical and general objectives of university degrees, the dualistic model still appears to exist at least on the level of officially stated objectives.

In an international comparison the Finnish university level engineering degrees have been considered master-level degrees and the polytechnic degrees bachelor-level degrees. The relationship between university level intermediate degree to be introduced in 2005 and the polytechnic degree is so far unclear.

3. Pedagogical training for engineering teachers

In this section we discuss the organisation of pedagogical training both in the university and the polytechnic sectors. We start by discussing some general trends that have given positive impact to raising the quality of education in technical universities.

3.1. Some general trends

Finnish universities have always been mostly research-orientated with the exception of some specific branches aiming at certain professions. This holds also for universities giving education in technology. Thus, there is a long tradition of scientific merits being the most important criteria for filling teaching positions. Especially this holds for professorships. However, winds of change have gradually entered the academic world during the 1990s. Starting from the University of Oulu in 1994 all Finnish universities have made a decision that teaching merits should also be evaluated properly when considering the applicants for a vacancy. In practice this means that all applicants for teaching positions should prepare a teaching portfolio, or an equivalent document, where they present their teaching experience and merits. Typically the portfolios include mostly factual information about the applicant's experience. However, he/she can also present his/her teaching philosophy. Most universities also require the applicant to give a public lecture, in which the teaching skills will be evaluated. Interviews are also commonly used to support the selection process. Still the selection process depends heavily on the universities and faculties, and in many cases the research merits are the dominating factors in the decision-making.

Typically the decisions to require portfolios from the applicants have also included a resolution to start portfolio training for the teachers. Such training activities have been carried out both nationally since late 1990s and also institutionally. At the same time more emphasis has been given also to provision of other pedagogical training for teachers. The forms and extent of training, however, varies a lot, from separate short courses to long-term pedagogical programs. We discuss these activities more closely in the next section.

Yet other activities, which have raised the status of education, are the evaluation processes carried out by the Finnish Higher Education Evaluation Council, FINHEEC. This organisation has evaluated over 200 curricula in Finnish universities and polytechnics during the past 10 years thus forcing the departments and faculties to self-evaluation and providing them also with international feedback on the teaching practices. An additional evaluation process is the selection of Centres of Excellence in Education. Every three years FINHEEC selects 20 units from the universities as Centres of Excellence in Education, based on applications submitted to the council by universities. The universities gaining such nominations get substantial financial support, which they can either give to the nominated unit, or use for some other purposes. This national level selection process has also induced processes for selecting local centres of excellence in education in some technical universities.

3.2. Pedagogical training in technical universities and faculties

Finnish law sets no requirements for the pedagogical training of university teachers. Thus most of the universities have set up systems and training schemes of their own. Here is the present situation in the three technical universities and three technical faculties.

Helsinki University of Technology (HUT) has a Teaching and Learning Development team (http://www.hut.fi/Yksikot/Opintotoimisto/Opetuki/-index.html), which has eight members, and which is the main responsible for most of the pedagogical training offered in HUT. Since 1999 HUT has offered its teachers a large programme (15 cr) on higher education pedagogy (see Section 2.4 for more details). Yearly, approximately 30 members of HUT staff participate in this programme, and due to growing demand, in 2004 two parallel courses will be arranged. In addition to the long course, shorter teaching skill related courses are provided for teaching assistants, and teaching staff of particular departments or laboratories. These are planned and executed in co-operation with the departments. Also general short-courses and workshops on various topics are arranged when needed. Teachers also have an opportunity to receive personal consultancy in various matters relating to their teaching. Training for the use of information and communication technology in teaching is provided in many levels and forms. Two large courses (TieVie II, 5 cr and TieVie III 10 cr) are arranged nationally in co-operation with other universities. These are complemented with local workshops and personal guidance. Theme-seminars and development days for the whole university are arranged a couple of times a year.

Lappeenranta University of Technology (LUT) has a Learning Centre (http://www.lut.fi/oppimiskeskus/Koulutukset/index.php) with currently seven employees. The first three credit course on pedagogy was arranged 2001. LUT started a large program (15 cr) on higher education pedagogy in 2002, and it has been conducted yearly since then. The programme consists of four mandatory courses and a personal development project. Courses can also be taken separately, according to personal needs. In the field of ICT in teaching LUT offers a wide range or different activities. Teachers can attend the large national TieVie-courses, but they are also offered various workshops and courses ranging from half to five days on different tools and methods. Personal consultancy is available too. Also a number of seminars for the whole university are arranged yearly.

Tampere University of Technology (TUT) is currently undergoing changes in the organisation of teaching development support, as the former Virtual University unit and Teaching Services team were united in summer 2003 to become the University Services unit. Before the reorganisation the Virtual University unit (2-4 people) arranged the training in the educational use of ICT and Teaching Services (4 people) the pedagogical training and other support functions. So far the staff in the Teaching Services has concentrated on administrational issues and TUT has not offered pedagogical training of its own. Instead the teachers have had the opportunity to participate in the Konnevesi-workshop (6 cr - see Section 2.4 for details), which is organised in co-operation with Universities of Oulu and Jyväskylä. TUT has a yearly quota for ten teachers in the workshop. The ICT in teaching –training was arranged along similar lines as those in other technical universities: possibility to participate in larger national courses and organisation of own workshops on various topics. TUT Virtual University unit had a strong emphasis on personal counselling and intensive work with the pilot projects.

Oulu University is a multi-disciplinary university with a technical faculty. It has a Teaching Development unit (http://www.oulu.fi/opetkeh/index.html) with approximately twenty people, which arranges the pedagogical training for the whole university. At the moment teachers of the whole university are offered 14 different courses (1-6 cr), supervision of work and possibility for departmental courses. The technical faculty, like other faculties, has a pedagogical co-ordinator of its own, who arranges faculty-specific courses together with the Teaching Development unit. Teachers from the technical faculty also have 5 yearly places in the Konnevesi-workshop. The national TieVie -courses are also on the reach of teachers and the national courses are in fact co-ordinated by the Teaching Development unit.

In Turku, where the Faculty of Engineering is situated, Åbo Akademi University (ÅA) has a Learning Centre (http://www.abo.fi/lc/), which is responsible for pedagogical support including the support for using ICT in teaching. There are nine people working part-time for the Learning Centre. Each year, the Faculty of Education arranges a 6 credit-course in higher education pedagogy. In addition to this, the teachers can take part in several short courses on various topics, arranged by the Learning Centre and the Centre for Continuing Education. Teachers can also attend the national TieVie - courses, aimed at improving the teachers' skills in and knowledge of ICT in teaching. In addition to courses, personal guidance and support, as well as extensive web-based support, are available in both general pedagogy and the use of ICT in teaching.

University of Vaasa has a joint Learning Centre (http://oppimiskeskus.tritonia.fi) with the local units of Åbo Akademi and Svenska Handelshögskolan. The centre offers training and counselling mainly in the area of ICT in teaching. There are currently fifteen employees working for the centre.

Finally we note that HUT Teaching and Learning Development team, LUT Learning Centre, ÅA Learning Centre and University of Vaasa Learning Centre also provide services for students.

3.3. Pedagogical training in Polytechnics

Unlike with universities, the legislation sets exact proficiency requirements for the permanent teachers in polytechnics. According to the Statute of polytechnic studies (Asetus ammattikorkeakouluopinnoista, 1995) the senior lecturers have to hold an applicable doctoral or licentiate degree and the lecturers a master's degree. In addition to this senior lecturers and lecturers teaching professional subjects have to have a minimum of three years work experience from the field. They also have to complete a pedagogical training worth of 35 study credits within three years from their nomination, if they do not already hold one when nominated.

To enable the fulfilment of these requirements there are five vocational teacher education colleges, which provide the 35-credit training for all the polytechnics. Basically the curricula of all the colleges consists of basic pedagogical studies, vocational pedagogical studies, teaching practice and other studies, but the realisation of the curricula varies a lot, with no standard components.

3.4. Examples of pedagogical training

Helsinki University of Technology was the first technical university in Finland to launch a larger pedagogical program for its faculty members. The program includes three pedagogical courses (2 credits each), a personal development project (6 credits) and 3 credits of other pedagogical courses, which the participants can choose freely. The courses deal with current pedagogical theories, classroom practices, use of ICT in education, and personal development as a teacher. The main focus of the programme is the personal project in which the teacher applies the new skills and theories to improve his/her own course, reflects on his/her experiences, and reports the results. The project plans, actions taken, and observations are discussed regularly within groups of participants and the training course teachers. A final report of 10-15 pages is written and the papers are published in the university report series.

The courses have been popular from the start and in 2003 almost 20 applicants could not fit into the course. Therefore new resources are allocated for the course next year. Most participants have submitted their application, because their colleagues have recommended the program, and they are thus strongly motivated from the beginning. The participants have reported that it has been very important to have a forum where they have an opportunity to discuss their experiences of teaching. Many people do not have this possibility in their own unit because of the tradition that teaching related issues are rarely subjects of discussion. The network of teachers has also remained after the people have graduated from the programme, and alumni meetings for all graduates have been organized yearly.

Another example of a long-term course of pedagogical training is the Konnevesi workshop of pedagogical expertise in technical sciences. This course (6 cr) is organized yearly in co-operation with the Universities of Oulu and Jyväskylä, and the Tampere University of Technology. About 20 participants from mainly these universities are admitted each year. The program has three face-to-face sessions during the year, each taking 2-3 days, where the participants come together in the Konnevesi research centre in the middle of Finland. The location in the countryside allows the participants to get rid of their daily work and concentrate on discussing and analysing their teaching traditions and courses in peace and quiet. Each participant designs his/her personal development project, implements it, and analyses and reports the results. Close co-operation with the other participants and the pedagogical tutoring experts is an important part of the program.

4. National projects and networks

In Finland, national networks and projects among universities are fairly common, as they still are small enough to remain functional. This section introduces four different networks, which operate on a national level.

4.1. PedaForum – network for university pedagogy
(https://tammi.oulu.fi/pls/pedaforum/pedaforum.in_English)

PedaForum was established in 1994 as a Finnish network for developing instruction and learning in higher education. From the beginning the network has operated on a sponsorship from Finnish Ministry of Education. Each Finnish university has a PedaForum contact person, and all of the network activities are intended for the wide audience of all university staff. Among the activities are:

- WWW-environment: descriptions and archives of the forum activities.
- Peda Online: an open e-mail list for discussions on participants.
- Workshops: intensive expert assisted development work on participant's own teaching.
- Teaching development conferences: national-level event for discussions and training organised annually by different universities.
- Newsletter on university pedagogy: a publication channel issued twice a year.
- Books: publication of research-based papers and descriptions of teachers' teaching experiments.
- Development teams: group work or groups working across departmental and disciplinary boundaries to examine pedagogical ideas and practices.

4.2. The Finnish Virtual University (http://www.virtuaaliyliopisto.fi/)

The Finnish Virtual University is a consortium of all Finnish universities. The consortium was founded in January 2002. The administration and projects of FVU are funded by the Finnish Ministry of Education, EU Structural Funds and the State Provincial Office of Southern Finland. The FVU services consist of the portal and different sub-projects. Sub-projects can be classified into four categories:

- Regional initiatives: co-operative projects among different universities for the development of virtual university activities.

- Joint initiatives supporting online education: service projects developing support services for online learning and education.

- Networks of academic disciplines: networks bringing together the best expertise from all over Finland in various disciplinary areas.

- Definition projects: production of reports of various topics in the field.

The portal is the main means of communication about different activities of FVU and also the 'home-base' of the online services developed in the projects and provided to all members of consortium. The online services include various web-based tools for teachers, students and administrators.

4.3. IT-Peda - network for use of ICT in teaching and research in universities (http://www.uta.fi/itpeda/index.html)

IT-peda network started its activities in autumn 1999. It aims at enhancing the use of ICT in teaching and research in universities, and the creation of support functions to these activities. IT-Peda is profiled as an expert network, and much of its activities are actually based on the networking of people and thus sharing the experiences and expertise of different people and universities. The network involves all Finnish universities and each university has one or more appointed contact persons, who meet 4-6 times a year. IT-Peda organises meetings for other experts in universities, too, and runs projects to develop services for the FVU portal.

4.4. TekPeda – pedagogical network for higher engineering education

The idea of establishment of a network to unite the educational developers in Finnish technical universities and faculties was born in autumn 2001 at Helsinki University of Technology. The Finnish Ministry of Education granted the funding for the start-up of the project for the year 2003, but the first meeting was held already in May 2002. The start-up project got into full speed in March 2003, the second meeting was held in June 2003, and the third meeting is planned for November 2003. The website for the network is currently under construction and is to be launched before the third meeting. All six technical universities and faculties have so far been involved with the network.

The basic objective of the TekPeda network is to develop the teaching in Finnish higher engineering education. Although the aim is quite similar to that of PedaForum, the central actors in Finnish technical universities felt that the field of engineering is so unique with respect to other university disciplines that it would benefit from a network of its own. TekPeda aims at creating processes where engineering education related knowledge is created, disseminated and applied in practice through training and consultancy provided for teachers. Another important function is the provision of social support and stimuli for the teachers, teacher trainers, and other developers of education.

The construction of the organisational and financial model for the network is still ongoing. However, already now it is evident that the network activities will have many levels. On one hand, the network will act a bit like IT-Peda and create an expert network through the meetings of the people most involved in teaching development of technical universities. In practice these people will probably be the employees of different centralised support and development units (see Section 2.2). On the other hand, a very important function of the network is to bring the teachers of engineering together across the institutional borders and provide them with forums to discuss the central and topical issues of engineering education. In this respect the activities will certainly resemble to some extent the PedaForum practices, with regular seminars and electronic discussion possibilities.

In the near future the focuses of the network are likely to relate to 1) the Bologna process of creating two-cycle degree structures, 2) co-operation in pedagogical training and 3) start-up of some research activities concerning the field of higher engineering education and pedagogy. As the universities are located quite far apart, the website and other electric means will be utilised in action as much as possible and convenient.

5. Conclusion

As a whole we can see that faculty development in Finnish technical universities has many different activities both in institutional and national level. The importance of building, supporting and developing pedagogical competences among the faculty members has been widely recognized, and the Ministry of Education has an important role in providing the financial basis for the work.

The process of pedagogical faculty development currently advances both top-down and bottom-up. With this we mean that the Ministry of Education and the governments of the technical universities have explicitly stated the importance of educational improvement. On the other hand, also many teachers have recognised the practical benefits of good pedagogical skills and thus started to educate themselves in pedagogy and apply new methods of teaching. They also disseminate the new way of thinking to their colleagues.

However, there are still big challenges to be met on the way to excellent engineering education. First, there are many faculty members, who do not recognise the pressures for changes or just ignore them. Second, the scarcity of resources together with the strong tradition of emphasising research work push many teachers to put all their effort to the research projects and neglect teaching. Third, the teaching culture is still strongly personal, and in very few laboratories teaching issues and methods are discussed openly. Instead each teacher takes care of his/her course on his/her own.

Thus, the road is long before we can introduce compulsory pedagogical studies for all university staff, but the time seems to be on faculty development's side as the younger teachers are more eager to participate in various available forms of pedagogical education. Probably this is partly due to their different view of good education compared with the view of older teacher generations, and partly because they consider pedagogical merits as advantageous in the competition for getting academic positions. We expect, however, that the process will take a long time, 10 to 20 years before a totally new educational culture is possible in Finnish technical universities.

(Section 2 is based on the chapter Engineering Education in Finland of the yet unpublished licentiate thesis of Johanna Naukkarinen).

References

Allt, S. (2003) *Kuvaajia koulutuksesta 2002*, [Charts about education 2002], Tekniikan Akateemisten liitto, Helsinki.

Asetus ammattikorkeakouluopinnoista (1995) [Statute of polytechnic studies], No 256.

Ministry of Education (2000) *Tuotantopainotteisen insinöörikoulutuksen kehittämistarpeet*, [The development needs of production focused engineering education]. Opetusministeriön työryhmien muistioita 7:2000.

Ministry of Education (2002a) *Tekniikan alan korkeakoulutuksen kehitysnäkymät. Selvitysmiehen raportti.* [Developmental view to higher engineering education. Investigators' report], Opetusministeriö, Helsinki.

Ministry of Education (2002b) Press release, October 31, 2002. Retrieved: August 19, 2003, from http://www.minedu.fi/opm/uutiset/archive-/2002/10/31_1.html.

Ministry of Education (2003a) *Taulukoita KOTA-tietokannasta 2002*. [Tables form KOTA-database (KOTA is a statistical database maintained by the Finnish Ministry of Education. It contains data describing university performance by institutions and by fields of study from 1981 onwards.) [2002]. Retrieved: August 19, 2003, from http://www.minedu.fi-/julkaisut/koulutus/2003/opm31/opm31.pdf.

Ministry of Education (2003b) *Ammattikorkeakoulut 2002,Taulukoita AMKOTA-tietokannasta*. [Polytechnics in 2002. Tables from AMKOTA-database](AMKOTA is a statistical database maintained by the Finnish Ministry of Education. It contains data describing polytechnic performance by institutions and by fields of study from 1997 onwards.). Retrieved: August 19, 2003, from http://www.minedu.fi/julkaisut-/koulutus/2003/opm32/opm32.pdf.

National Board of Education (2003) *Tutkinnot ja niihin johtavat koulutusohjelmat*. [Degrees and respective degree programmes]. Retrieved: August 19, 2003, from http://www.oph.fi/koulutusoppaat/amkopinnot/pdf/koulutusohjelmat.pdf.

Orelma, A. (1996) *Insinöörikoulutus epävarmuuden yhteiskunnassa*. [Engineering education in the risk society]. Turun yliopisto. Koulutussosiologian tutkimuskeskus Raportti 36.

Teknillisten tieteiden yhteisvalintaopas (2003) [The admission guide to higher engineering education 2003]. Retrieved: August 19, 2003, from http://www.hut.fi/Abi/haku/valintaopas2003.pdf.

Chapter 9

The Educational System within Engineering in Norway – Development Strategies at Faculty Level

Margrete Fuglem

1. Introduction

Engineering education at graduate level (Master) and at undergraduate level (Bachelor) has gone through great organizational changes in Norway. The Norwegian Institute of Technology (NTH) and the colleges of engineering are no longer institutions of their own, but now they are parts of either the universities or the 18 regional (State) colleges/university colleges. Concurrently, ongoing discussions concerning the quality of teaching and learning have developed the programmes of study considerably. The strategies to stimulate developments at faculty level have comprised central measures and recommendations in connection with the request for more local engagement.

The current article addresses firstly the organizational changes and secondly the strategies and measures within engineering education which influence faculty development.

2. The structure of the engineering education

Norway has four universities. The Norwegian University of Science and Technology (NTNU) in Trondheim holds main responsibility in the country for graduate- and doctoral-level technological education. NTNU enrol approx. 19 000 students, and almost 7500 are graduating engineering students, which is about 80 % of the total number of graduating engineering students in Norway. The three-year engineering education (Bachelor) is offered at 18 regional colleges and was established in year 2000. The joint declaration of the European Ministers of Education, the Bologna Declaration from June 1999 and the Bologna Process have influenced the organization of engineering education in Norway.

Admission to engineering studies is from upper secondary school, General Studies or Vocational studies: First level (upper) Mathematics and second level Physics are mandatory subjects from school. For students without the required courses, different preparatory courses are offered.

    A. Main structure:

- 3 years Bachelor
- 3 years study pivot
- 2 years Master
- 5 years integrated Master
- 3 years PhD

    B. Colleges of Engineering offer:

- 3 years Bachelor, professional, broad number of programmes
- 3 years Bachelor, professional, special programmes for continuation into a Master programme at NTNU, in a broad number of programmes
-

    C. University Colleges of Engineering offer:

- 3 years Bachelor, professional, broad number of programmes
- 3 years Bachelor, professional, special programmes for continuation into a Master programme on NTNU, broad number of programmes
- 2 years Master based on a Bachelor within the same profession. Few programmes
- 3 years PhD, often with an additional year as Assist. Prof.(2% of total) Few programmes

D. General Universities and the Agricultural University of Norway offer:

- 3 years Bachelor. Very few programmes and very few candidates.
- 2 years Master based on a bachelor within the same area. Very few programmes and very few candidates.
- 3 years PhD. Very few programmes and almost no candidates.

3. Recruitment, admission and production

At NTNU, admissions of the engineering students are registered at four different faculties:

- Faculty of Engineering Science and Technology
- Faculty of Natural Science and Technology
- Faculty of Information Technology, Mathematics and Electrical Engineering
- Faculty of Social Science and Technology Management

Approx. 1 500 engineering students are admitted per year. 27% of these are women.

Studies/programmes offered at NTNU with number of students from 2002:

Energy and environment – 120 students
Electronics - 92 students
Engineering cybernetics – 100 students
Communication technology – 111 students
Computer technology – 193 students
Chemistry – 95 students
Marine technology – 74 students
Mechanical engineering and (product development) – 120 students
Civil engineering – 141 students
Physics/Mathematics – 114 students
Industrial economics –117 students
Technical design – 20 students
Engineering science and ICT – 60 students
Materials technology – 32 students
Product development - 116 students

In 2002, a total of 2684 students started their engineering studies at the regional colleges in 2002. Among them, 455 (17%) were women. At the same time, 1925 students graduated. The programmes of study are organized as the traditional courses within civil engineering, such as computer science, chemistry and mechanical engineering. Some of the new study programmes do not fit into these categories, for example studies in environment and management.

3.1 Recruitment strategies

The situation during the last few years has brought about a more conscious strategy concerning recruitment. Successful recruitment of the right number of competent undergraduates is important. During recent years, the amount of students in secondary school who studies first level Mathematics and Physics, the motivation and the opportunities for young people to enrol into a science and engineering study have decreased. As a consequence of this, the need for recruitment activities and for preparatory courses has increased. Altogether, this is a common challenge for both the Norwegian University of Science and Technology and to the regional colleges. Recruitment and admission in 2002, at NTNU have been evaluated as being satisfactory, because most faculties have experienced enrolment of as many undergraduates as they had hoped for. Very few faculties chose to reduce the number of students rather than lowering the minimum admission requirements.

At the regional colleges, they have met many of the same challenges in relation to recruitment. Despite an expressed shortage of engineers in Norway with specific types of bachelor's degrees, it seems however difficult to obtain a sufficient number of graduates.

The national parliament (Stortinget) is aware of this negative development and has granted a considerable amount of money to support preparatory one-year courses, at the regional colleges.

"Women in engineering" was a slogan for some years. In the article "The Strategic Foundation for Norwegian University of Science and Technology" from 1996 (Strategic foundation for Norwegian University of Science and Technology, 1996) it was stated:

It is clear that women often interpret and handle information differently from men. Since technological studies for a long time have been dominated by men, it seems very important to take advantage of the different experiences of women where disciplines are developed, as well as in the management and governing bodies of the University.

And further:
NTNU is to actively work in recruitment to provide NTNU with the best i human resources, and particularly improve the recruitment of females in education, research and management

The focus of staff development is of outstanding importance in order to make engineering an area, where women are better represented. However, to recruit female teachers will take time. At NTNU in 1995, 13% of all teachers were women (architect teachers included). In 2000 the number was 9.5% and in 2002 a small increase were noticed, 17.5 %.

In all the engineering studies at NTNU in 2002, a total of 402 women graduated and they count 27% of all the undergraduates. The number of women participating in engineering studies at NTNU differs greatly between the departments – from the Department of Chemical Engineering, who has 60% women, to the Department of Marine Technology who has 13.5% (Rapport opptak, 2002).

Measures have been established to stimulate a higher proportion of women in engineering studies. Computer science is a developing discipline which has undergone rapid growth, but the number of women within this programme has been much lower than hoped for. In 1992, the electrical engineering and computer science faculty only had 9.7% female undergraduates. In that relation, NTNU started to award female undergraduates with extra points for studying computer science. As a result, a greater representation of females was interested in the subject. Hence, in 2002, 23.8% women participated in the programme. After that experience, all engineering departments – apart from chemical engineering and industrial economics - award women with two additional points (normal points required for admission is about 50). This adjustment in measures seems to cause no problems.

3.2. Results

In 1997, 1401 students started their five-year engineering programme at NTNU. In 2002, 1012 graduated at Master of Science level (Norwegian name of degree: Master i Teknologi), meaning that 389 students have either given up on their studies or they have decided that they will need more time to complete their studies. In 2002, the failure percentage for all engineering programmes was 11.87 % (www.nsd.uib.no).

A frequently used concept concerning quality matters in higher education is *production,* as the central authorities often use this concept as a quality criterion. When in 1997, 1401 undergraduates started their engineering studies at NTNU and 1012 engineers graduated after 5 years, it can be concluded that 72% have completed their studies within the scheduled time. Comparing these figures as a specific expression of quality, would be a rather incorrect assessment, because some undergraduates spend a year or two to study abroad, others have a formal leave of absence. Again others may doubt that this career path is the right one for them and drop out. However – it is a good thing to watch the pattern of the undergraduates and their progression in the study system. The regional colleges have a curriculum with more central control, whereas mathematics and physics still have to remain subjects of their own with separate exams. As a result of this, more and more colleges struggle with high failure rates within these courses.

4. Quality development strategies - Staff development.

Already in its 70th year, The Norwegian Institute of Technology offered pedagogical courses to their teachers. In 1982, The Board of Professors decided that all new staff members should follow a pedagogical course. However, it would be a rather incorrect assessment to claim that the course was compulsory from that year on. But from 1988, an educational centre with three and a half person was established and NTH developed a better strategy to provide new staff members with teaching skills. In one of the best years, 1992, 45 teachers (professors and associated professors) completed the course PEDUP (The Pedagogical Developmental Programme for new staff members). The title changed from "course" to "programme" (Fuglem, 1991). By observing the structure and content of the programme during a period of some years, it became clear that it was a development from teacher training to a programme focusing on both students' learning, teachers consciousness in his/her own teaching process, motivation and also personal development. Furthermore, this was an adaptation to the new policy of the institution (NTH) who stressed that a more various teaching practice was important and who focused much on the student's future competences as an engineer. The courses only included participants from the engineering education, and it was considered appropriate if more than one person from each department participated in the same course – in order to stimulate what could happen "on home ground" - not least after the programme was finished.

NTNU was established in 1996 which caused a reorganization of the staff development programme. The policy was "integration" and the Teacher Training Programme was made responsible for teaching staff development at

NTNU, which meant that people with expertise within education of school teachers were regarded as having the sufficient competences for staff development within all programmes at university, such as engineering and medicine. Afterwards, the programme for new teaching staff included participants from the entire institution and from that year on, alternative plans were no longer relevant; such as establishing a Centre (or Institute) for Learning and Teaching in Higher Education and the concept of "engineering education". In Europe, however, the trend is to establish university centres to assist their own institutions (Kolmos et al, 2001). Bowden and Marton, who are concerned with the definition and development of professional competences, recommend more differentiated programmes for staff development. In that connection, they state that for some years the acts and processes of knowledge have become separated from knowledge itself and furthermore, they argue for a reunion. Learning in the sense of formation of knowledge, should be a specialization within every domain of knowledge (Bowden & Marton, 1998). Consequently, goals (professional competence) and the content of the engineering studies should be of great importance to the way in which teaching and learning are organised for the engineering students. The interdisciplinary project with experts in team, which are compulsory for all engineering students in the 8th semester.

The engineering programmes at NTNU have gone through a great process of change. In that connection, it would be relevant to ask: how the teachers are prepared for these changes? And which staff development measures have been established?

In their summery report, The Will to Improve, the (Curriculum Committee, 1993) took it for granted that the engineering teachers would now be offered several important pedagogical courses. Hence, after having established NTNU in 1996, a pedagogical week for all the teachers was arranged in September 2000 with seminars concerning new learning methods and exams. The participation rate was disappointing. Five seminars had an average of 56 participants, and most of these people were extraordinary motivated to participate. Therefore, it is necessary to ask a few questions concerning the strategy: Is it interesting and motivating for teachers to participate in seminars that raise questions of a general character, in order for them to suite a large scale of disciplines? If the presented knowledge do not enable the participants to make clear links between the instructional ideas and how to use them in their own teaching, will the teachers find it worthwhile to spend their busy time on seminars? Finally – do the teachers already adapt to the new ideas which recommend changes in the current teaching practice?

When handling specific tasks such as being a teacher in an Interdisciplinary Project/Experts in teams, the teachers have to complete a compulsory course for this particular purpose. The focus here is related to being an Expert in a team. What is going on in the groups? In addition, a psychologist is responsible for this course and by evaluating the course, it becomes clear that the task of making such a course acceptable and relevant for teachers in the engineering department is not an easy one (Fuglem, 2002).

For teachers who teach in interdisciplinary projects, there are regular meetings between the active teachers in order to exchange experiences and discus strategies for actual challenges. These meetings are not compulsory, however, the strategy to listen and to learn from teachers who are not educational professionals, but who are active practitioners - is widespread and optimistic. The effect, however, is mostly in terms of motivation, which should not be under-estimated.

At the regional colleges, each college is responsible for deciding the measures for staff development. Some of the institutions arrange compulsory programmes for their teachers and set a time limit of three years for their tenures/full-time employees to complete it.

5. *The new five-year engineering degree*

The Norwegian University of Science and Technology (NTNU) was founded on 1st January 1996 by a governmental decision to reorganize The University of Trondheim. The university included The Norwegian Institute of Technology (NTH) together with the faculties of Social Science, Humanities and Medicine.

The creation of NTNU aimed at strengthening science and technology and at the same time to enhance the interaction with non-technical disciplines towards a more integrated approach to problem solving. Accordingly, interdisciplinarity is a concept with specific relevance to science and to education as well.

The engineering education has been influenced by the historical event, such as the creation of the new university. Though in 1993, the earlier Norwegian Institute of Technology (NTH) did already initiate a process to improve the engineering studies by establishing a Curriculum Development Committee. This committee made conclusions which did in many ways support the establishment of NTNU. The urgent need for interaction between science and technology on one hand and social science and humanities on the other was increasing. The Curriculum Committee concluded in their summery report "*The will to improve*" that the role of engineering and the role of the engineers in society is about to change. Engineers can no longer focus on technical solutions and problems only, but they have to be increasingly aware of the totality in problem definition and alternative solutions, as well as their social and ecological consequences.

The committee proposed that The University of Trondheim (its name in 1993) should remain the national responsibility for the engineering educations at higher levels. The main aim of the engineering studies should be a competent and skilled graduate – competent for the future challenges as an engineer. It may be referred to as sensational that The Curriculum Committee pronounced that teaching and learning methods should be the guarantee for quality in the study programmes, which meant that not only the content was important, but more attention should be paid to the way in which students work at university level - the learning methods. During the last years, it has been seen that the curriculum was heavily overloaded and more and more fragmented, including many exams. A solution to this problem would be to reduce the number of courses and the number of exams. Lecturing should still be the overall teaching methods, but they should be reduced in order to benefit from the development of more student-active methods, such as project and problem-based learning.

Experts in team (EiT) and interdisciplinary projects, held a unique position. This course was made compulsory for all students at 8th semester, and a group of three (a professor in engineering, an educator and a student) was established in 1997 and were made responsible for the development of EiT. During three years of designing, experiencing and reporting to the board of engineering education, the course was considered ready to provide in year 2000. Approximately 40 teachers are now involved in EiT each year. A great challenge is to create a case which can function as an interdisciplinary project for students from different engineering programmes. Further - to encourage the launch of innovations and stimulate new development, the board of engineering education (GUS) has awarded financial resources to measures concerning the implementation of the new 5-year curricula. The framework established for these measures was wide, but there was specific focus on the *Specialization* in the 9th semester and the change to a pedagogical shift towards more ICT in teaching. The subject for specialization in the 9th semester was a considerable challenge to the teachers when concerning their will to integrate their subjects and collaborate with colleagues. In this period, the Stortinget allocated about NOK 5 million to the engineering education at NTNU.

The new 5-year degree had great impact both on pedagogical matters and on the organization of students. On the basis of these purposes, new ways of organizing became necessary. Each faculty was appointed a *coordinator for the academic year* - a measure to avoid problems and chaos, when more projects and a less structured curriculum for the students were to be expected.

From year 2003, the undergraduates will be invited to take an integrated course in order to brush up their mathematics and to strengthen their ability to cooperate. The professional focus from *"The will to improve"* stresses the importance of this ability for engineering students and for staff members to strengthen the links between academics and relevant members of engineering companies.

At the time being - the new 5-year programme in engineering is about to fulfil its implementation – which means that the new curricula is carried through in the entire study programme. Nevertheless – more measures need continuous follow up. Furthermore, the main challenges in a radical project such as the new programme are the way from VISION, through STRATEGIES to REALIZATION. It seems that visions/goals are clear (to develop a relevant competence for the graduate engineers) but the strategies and "ways" are weaker. The renewal of the engineering education at NTNU is initiated through central decisions, but its realization depends mainly on, how "the players of the field", such as the teachers and the students can meet the challenge and create the pedagogical measures. For this to succeed, we require a conscious strategy for teaching staff development.

From the 1980s new engineering teaching staff at NTH were obliged to join pedagogical courses. More than 200 teachers were informed and trained for more various pedagogical practices in these courses. It was a pleasure to observe that many of the teachers were front runners within development agency. However, will the system concerning pedagogical education of new staff be sufficient today, when all teachers have to meet the demands of creativity and pedagogical know-how in order to fulfil upcoming expectations?

6. The Quality Reform in Higher Education

In June 2002 the Norwegian parliament gave its assent to a comprehensive reform for Higher Education in Norway: The Quality Reform. This reform was influenced by the more common tendencies concerning development of Higher Education in the western world. During the last decade or so, universities around the world have been subject to increasing pressure to become accountable and more efficient. At the same time they have had to adapt to the new challenges as being mass universities. In addition, the community is demanding more explicit accountability from universities and academics, and the institutions now have to document the way they address the demand for quality in teaching and learning. An approach has been to establish systems for quality development and quality assurance. The Quality Reform in Norway in principle holds the institutions responsible for developing their own quality improvements and their own quality systems.

Most relevant attempts within the professions have been to define clearly the expected outcomes of learning, such as the competences. The new five-year engineering degree at NTNU describes the skills which are required from engineers in the future.

To implement The Quality Reform at NTNU, three committees and five learning areas were established as a faculty development strategy. The expectation was that teachers from all study programmes at NTNU could meet in the areas and learn from each other. The Teacher Training Institute was expected to provide guidance and recommendations to the faculties and to the teachers. The five learning areas were: ICT in Learning, Learning Goals and Assessment, The Pedagogy of Extensive Courses, Guidance of Groups and use of Teaching Assistants, Problem-Based Learning activities.

7. Into the future

The "Engineering education" in other European countries seems to be a strong concept, and the establishment of the UNESCO international engineering centre in Aalborg from 2002 manifests an international orientation as well. Strategies and measures concerning engineering education in Norway seem to develop into another direction. Organizationally, integration encourages more common strategies concerning faculty development. This may be a good and necessary development, but it complicates the current picture. A preliminary strategic plan for all study programmes at NTNU from 2000 (Draft for teaching strategies, 2002) recommends:

- That all faculties are responsible for perpetuating and further developing the quality of pedagogy within the different disciplines.
- The professional educators (university pedagogy) have to be active in consulting and evaluating this activity.

Though further questions still have to be raised at NTNU:

- Do we have the proper organization for such a pedagogical challenge?
- Are the measures functional, which can make this vision become reality?

- Is staff development making teachers more competent in order for them to provide a conscious and varied pedagogical praxis which is a preconception for successful faculty development?
- How can teaching and learning in the institution be a matter of research? (For instance, how are the recommendations from the strategic committee reported and followed up?)
- How can the status of pedagogical development be afforded?
- How can ambitious and quality-conscious teachers be rewarded?
- Are university educators competent enough to offer this demanding assistance to faculty and staff (what skills do they need)?

References

Bowden, J. & Marton, F. (1998) *The University of Learning*, Kogan Press, Gøteborg.

Database for statistisk høyere utdanning. Retrieved July 6, 2004 from http://www.nsd.uib.no.

Fuglem, M. (1991) *Pedagogisk utviklingsprogramme for nyansatte* (PEDUP), Ved NTH. Tidsskriftet UNIPED, No.3.

Fuglem, M. (2002) *Rapport nr. 3 om Innføring av Eksperter i team – tverrfaglig project" NTNU.*

Kolmos, A., Rump, C., Ingemarson, I., et al.(2001) Organization of staff development – strategies and experiences. European, *Journal of Engineering Education*, Vol.26, No.4, pp 329 – 342.

Rapport opptak (2002) *Allmennvitenskapelige studier*, Sivilingeniør/sivilarkitektstudiene, NTNU.

Strategic foundation for Norwegian University of Science and Technology, (1996).

The Curriculum Comitèe, Summary Report, *Will to improve* (1993) NTNU.

Utkast til undervisningsstrategi for NTNU (2000) (Draft for teaching strategies).

Chapter 10

Faculty Development strategies at the Danish Enginering Education

Anette Kolmos and Ole Vinther

1. Introduction

Engineering education is a profession based education with strong relation to society, especially industry. Therefore the implementation of employability into engineering education during the last 20 years, has not caused the same challenges as in many other subject areas. However, still Danish engineering educations have undergone a lot of changes.

In this article, the present structure of engineering education in Denmark, faculty development centres, and faculty development strategies at the universities and university colleges are shortly described in order to give a short description of the engineering context.

2. Engineering education

In Denmark, engineering education is offered at two levels. Bachelor degree, called "diplomingeniør", with a duration of 3½ years, and a Master degree, called "civilingeniør", with a duration of 5 years. In order to adapt engineering educations to the Bologna declaration, the structure of engineering educations will be changed to 3 + 2 years. This change takes place at the moment and has a lot of implications for setting up new curriculums.

In Denmark there are 3 universities, offering bachelor and master degrees, and 5 university colleges for engineering, offering bachelor degrees.

The 3 universities providing engineering educations are:
Technical University of Denmark (www.dtu.dk),
Aalborg University (www.aau.dk),
University of Southern Denmark (www.sdu.dk).

The 5 university colleges for engineering are:
Copenhagen University College of Engineering (www.ihk.dk),
Odense University College of Engineering (www.iot.dk),
University College of Aarhus (www.iha.dk),
Herning Institute of Business Administration and Technology (www.hih.dk)
Vitus Bering Denmark in Horsens (www.vitusbering.dk).

There is some pressure from the Ministry of Science, Technology and Innovation and the Ministry of Education in order to motivate the smaller institutions to merge with other educational institutions. During the last 5 years, this has resulted in new co-operative models between the universities and the university colleges where the universities provide research based education whereas university colleges provide research related education.

There is one professional association for engineers in Denmark, IDA- The Society for Danish Engineers (www.ida.dk). IDA plays a very important role for creation of the professional identity for engineers. As indicated by having an engineering profession, a special culture and self understanding of the role of engineers have been developed by weekly newspapers and well organised activities.

There are about 17.000 engineering students and approximately 2550 members of staff at the institutions offering engineering educations. Furthermore, engineers teach at several other bachelor level technical educations.

3. The academic career

At universities in Denmark, the career of academic staff takes the following path.

PhD scholarship, with duration of 3 years. The PhD programme consists of 30 ECTS course work, 120 ECTS for writing a PhD-thesis, and, finally, most PhD-students have a contract for a teaching load of 30 ECTS.

Assistant professor, with duration of 3-4 years. During this period the assistant professor must join pedagogical training.

Associate professor, which is a permanent position, and professor, which is either a permanent position or a fixed-term appointment.

To get a promotion, there must be an advertised position, which everybody can apply for. It is not possible to get a promotion within the same position as in the British system.

To become associate professor, one must document pedagogical competencies as well as research within the subject area and research competencies.

At university colleges, the career of academic staff takes the following path. Assistant professor, with duration of 3-4 years and associate professor, which is an advertised position that requires a minimum of 100 hours training in pedagogical issues or equivalent experience. There is no requirement for research.

4. Dominating pedagogical methods

The traditional teaching model with classroom teaching and a single project in the end is under pressure (figure 1).

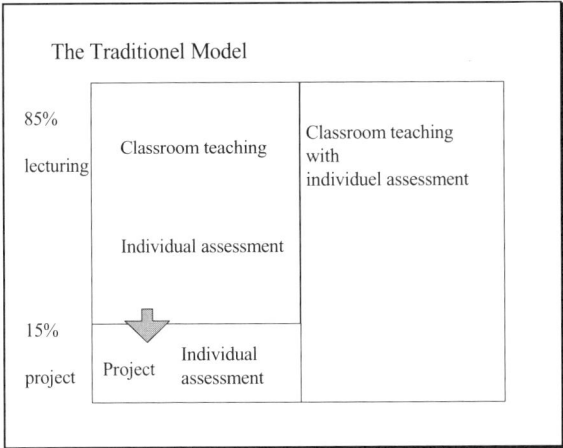

Figure 1 The traditional model and the Aalborg Model

Most engineering schools are now changing to a much more project organised curriculum. They develop their own models – highly inspired by the Aalborg model described in the figure above. The dominating pedagogical model is shifting from a discipline oriented concept to a problem based and project organised curriculum.

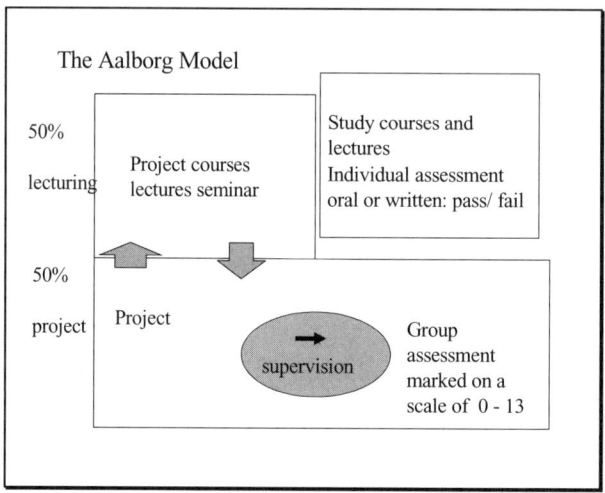

Figure 2 The Aalborg PBL model

As shown in Figure 1 and 2, the structures of the traditional and the new programmes are fundamentally different.

The Aalborg Model is built on problem-based project work in which approximately one half of the students' time is spent on group project work, whereas the other half is spent on traditional lectures. All project work is done in teams, and the same model is followed from the 1st semester until completion of a masters' degree. During the span of the university degree programme, the groups normally become smaller, with typically 6-7 students in the 1st year to approximately 2-3 students in the final semester.

Two types of courses are offered to students at Aalborg University - those that are directly related to the students' project work and those that have nothing to do with the project work in terms of subject matter. In the former type of course, students are assessed as a group through evaluation of their common project work. Approximately ¾ of the students' total study time in these courses is tied to the project work itself. Only in the last-mentioned courses, (i.e. those in which the subject matter is not related to the project) students are assessed individually.

At the university colleges one will find a variety of combination between course work and projects, the degree of problem orientation and openness, the length and duration of the projects, the size of the teams and how they are established, the role of the supervisor and not at least where in the educational progression project and problem based learning is implemented.

5. Pedagogical training - National network and institutional centres

In Denmark, institutions offering engineering educations have formed a national pedagogical network, called IPN (Danish Pedagogical Network for Engineering Education). IPN consists of all engineering institutions in Denmark, both universities and university colleges of engineering. The Ministry of Education's "quality improvement" pool financed the network for a period of 7 years, ending in 2003. Now the institutions have taken over the financial responsibility.

The purpose of the network is to improve the quality of engineering education by pedagogical and didactical activities:

- Inspiration to pedagogical and curriculum development activities, and initiate and coordinate such activities.

- Providing training and education within the field for part time teachers, PhD students, assistant professors, associate professors and professors.

- Collecting and disseminating information concerning pedagogy.

- Initiating curriculum development projects at the individual institutions.

- Creating a forum for the exchange of ideas and experience at an institutional, national and international level.

All member institutions have a staff member who is employed to work part-time for IPN. Additionally, a head of the organisation is employed full-time.

At Aalborg University, a faculty development function was established during the early 90'ties (Kolmos et al., 2001). The Technical University of Denmark had a similar centre, called CDM, which has recently been reduced in resources and changed organisational function. The national network and the two centres have been complementing each other.

Centre for University Teaching and Learning (PUC) (http://www.puc.auc.dk), Aalborg University, was established in 1995 for a three-year period and was confirmed in 1998. Aalborg University and external project funding finance the centre. The university has about 13,000 students at three faculties (humanities, social science and engineering). Recently, a new UNESCO International Centre for Engineering Education, Centre for Problem-based Learning (UCPBL) has been established. UCPBL has primarily external/global activities; however, there is a close cooperation internally.

6. Formal structures and requirements

In Denmark, the requirements for pedagogical training of the academic staff at universities and university colleges are very different.

At universities, pedagogical training is mandatory. The Ministry of Education has specified that all assistant professors (a three-year position after the PhD level) must receive pedagogical supervision and tutoring and must receive a written statement regarding their teaching qualifications (Position Structure Briefing, 15 March 2000).

The individual institution has to fulfil the requirements, which primarily is done by scheduling compulsory courses in which the participants contribute approximately 200 hours. Assistant professors must document teaching abilities (including the written statement regarding teaching qualifications) in order to qualify for permanent tenure. The most recent revised regulations (adopted in 2000) specify that the teaching qualifications should be given more weight than in previous years.

The primary strategy for universities in Denmark is the compulsory courses to the assistant professors. Prior to this requirement, it was in fact quite difficult to motivate university teachers, actively involved in research, to participate in longer pedagogic training courses. However, teaching assistants and PhD-students are motivated to participate in the shorter courses in order to learn the technical pedagogic principles.

There are no formal requirements for the university colleges, but in 1997 IPN was established primarily to train the staff at the university colleges. So even though there are no formal requirements, the rectors of the engineering institutions have initiated that all members of staff at institutions offering engineering education must document pedagogical training of at least 10 ECTS.

7. Training activities

The national pedagogical network and the two institutional centres are running a lot of different activities, as mentioned below:

Certification courses that are courses the participants can register for individually. The certifications courses are mostly mandatory courses at universities which have a somewhat similar structure even though the individual elements are given different names. However, the contents of the CDM courses and the PUC workshops reflect to a certain extent the differences in teaching cultures at the two universities. Technical University of Denmark (host of CDM) is an older university with a long tradition of courses based on lectures and exercises, whereas Aalborg University (host of PUC) is a quite new university with a problem-based and project organised curriculum.

Involving both professional educators and professors from the participants own departments in the supervisory aspects of the programme has proven highly beneficial for both the assistant professors and the professors serving as supervisors. The assistant professors have someone within their department with whom they may discuss their activities and the supervisors become involved in the university didactics and often develop a personal commitment to the training programme. Thus, this method of organising the course has served to involve experienced professors in pedagogical training.

Courses held on the basis of an inquiry from departments or from a group of teachers are other kinds of courses. These types of courses give a group of colleagues the same language and there is a possibility that the team develops the educations.

Participation in courses does not appeal to experienced teachers mainly because of lack of time but also because they do not want to regard themselves as students. Therefore it is important to arrange other types of activities where they can use their experiences in another way. Arranging seminars and conferences is a way to meet this requirement.

Consultancy and curriculum development projects are strategies which meet both experienced teachers' requirements and also contain a great deal of implementation within the organisation.

Finally, information is provided through many, different kinds of activities, such as newsletters or short lectures about pedagogical issues.

8. Reflections on the Danish organisation

The Danish IPN has been incredibly rewarding. It ensures a high degree of exchange of information and, especially, it inspires the Danish engineering educations to further pedagogical development. However, it is beyond doubt that if serious pedagogical improvements and implementation shall take place then institutional centres must be established. The expert knowledge must be present at the institutions; it is not enough to participate in national courses.

There is a considerably difference between the institutional centres at the universities and the national pedagogical network. The universities are research based and therefore the faculty development units are research based as well. This makes it possible to develop all activities further, not only by experience but also by an evidence-based approach and by more theoretical research.

However, it is important that the staff members possess both subject knowledge and pedagogical knowledge. Otherwise, there is a risk that these faculty development units are regarded as a group of theoretical experts concerned only with research and therefore useless to professors in engineering. To gain respect, it is important that staff can "speak the language"- in other words are dual qualified in terms of both knowledge of pedagogy and the particular subject field in engineering.

References

Kolmos, A., Rump, C., Ingemarsson, I. et al. (2001) Organization of Staff Development - Strategies and Experiences, *European Journal of Engineeng Education*, Vol. 26, No. 4. pp 329 - 342.

The Ministerial Circular on Job Structure for Academic Staff working with Research and Teaching at Universities, etc. under the Ministry of Research and Information Technology (2000).

www.puc.auc.dk, Retrieved July 7, 2004.

www.ipn.dk, Retrieved July 7, 2004.

www.iti.auc.dk, Retrieved July 7, 2004.

Vinther, O. & Kolmos, A. (2002) National strategies for staff and faculty development in engineering education, *Global Journal of Engineering Education*, Vol. 6. No. 2.

Chapter 11

Reflection for Staff and Faculty Development

Säde-Pirkko Nissilä

1. Abstract

The changes in learning conceptions necessitate a new orientation to teaching. It is important to make the teachers´ and teacher trainee's silent knowledge, their implicit conceptions explicit, and to facilitate their ability to learn from experience and theory through systematic reflection. Since new understandings are created both on personal and collective levels, they are very individual on one hand, and very collective on the other.

It is essential for professional development to stress metacognitive knowledge and skills, self-understanding and self-monitoring as well as understanding the phenomena and situations in the classrooms. This became evident from the first analysis in this article.

Increasing pedagogical content knowledge and subject matter skills as well requires courage to encounter problems, solve them and evaluate personal action. An asset throughout this process is the skill of systematic reflection which was shown in the second analysis. Self-assessment e.g. through portfolio work creates understanding of the qualitative contents of teaching and thus makes the perspective wider. These kind of reflective practices are possible both in extensive teacher education programs for young adults and in shorter programs for adult experts.

The awareness of teacher trainees´ use of their own experiences and the consistence promotion of critical writing and thinking skills are supposed to strengthen the awareness of different aspects of education. Reflection is the key to understanding oneself as a teacher, the learner perspectives and the practical actions in teaching with their possible consequences. Teachers are expected to have good communication skills, specialists in team work, ready to take initiative, to foster learner autonomy and master their subject matters.

2. Reflection and new professionalism in teaching

Researches have attempted to investigate the relationship between teachers' cognition and their behaviour (Calderhead, 1996; Richardson, Anders, Tidwell & Lloyd 1991; Meijer, Beijaard & Verloop, 2002). Teachers' beliefs usually reflect the actual nature of their practice, and problems arise in making teachers' beliefs explicit (Kagan, 1992). A reason for the problems might be that teachers cognition and classroom action often are studied separately. Schön introduced the concepts knowing-in-action and reflection-on-action, indicating how teachers think while they teach: "Our knowing is ordinarily tacit, implicit in our patterns of action and in our feel for the stuff with which we are dealing" (Schön, 1983).

Many of the recent studies have tried to reveal the areas teachers think of when they reflect on their work. Schulman's reports (1987) have inspired many others to investigate the kinds of knowledge that underlie teachers' actions. An important outcome is the development of the concept "pedagogical content knowledge". It encompasses understanding of common learning difficulties and preconceptions of students (Van Driel, 1998). It is believed to be an essential domain because it focuses explicitly on the knowledge and skills that are unique to the teaching profession.

Pedagogical content knowledge can be divided into the sub-categories of subject matter knowledge, knowledge of general pedagogy and knowledge of learners and specific learning contexts. Thus it is a unique domain of knowledge possessed by teachers. It can be seen as teachers' internalised procedural knowledge of how particular topics, problems and issues can be adapted, organised and presented to given learners with diverse interests and capacities. (Shulman, 1987.)

2.1. Conceptions of knowledge and learning

The ways the teachers conceptualise their professional tasks depend on their basic philosophical orientation and their conceptions of man, knowledge and learning. The ways the teachers see themselves as human beings, learners and professionals influence their pedagogical choices. Therefore the teachers need to be in constant process of clarifying their educational philosophy and values and the ways they organise their work.

An essential distinction of the conceptions of learning is how active a role the learner (teacher and student can both be learners) is playing in the learning process. (The polar ends are the role as an active constructor of meanings and as a passive recipient of information.) In the modern conception of learning the learner's inner activity and initiative are considered as essential as his/her social environment.

Thus changing meanings is an active process. It means active involvement in sharing and understanding. It also means negotiating about aims, contents and learning. An active learner is like a constructor of a road map including both learning goals, tasks, learning processes and oneself as a learner.

Consequently personal growth can be defined as changes over time in behaviour, knowledge, images, beliefs, or perceptions. On the other hand, unless teachers develop consciously their thinking and doing, their personal beliefs remain usually unchanged and appear in the classroom practice (Zeichner & Gore, 1990). Autonomy and responsibility are aims for teacher development and education for citizenship. If teachers are expected to strive for those goals, they should have opportunities to internalise them experientially as part of their pre-service and in-service education. Autonomy, which is a prerequisite for the teacher's own professional growth, also aims to promote responsible student learning and school development.

2.1 Autonomy, responsibility and collaboration in teaching

The new role of a teacher as a negotiator, an organiser of learning situations and a member of the teaching staff leads to new goals in professional development, i.e. collaboration, participation and interaction entailing also a stronger obligation towards colleagues in the context of faculty development. Autonomy in a learning organisation, i.e. among the colleagues or the staff, can best be described as partnership, which facilitates various networks of the staff.

To become autonomous an individual must have a range of choices, she/he must know them and know also that they are open for his preferences. Promoting this kind of awareness is one task of staff development. To become autonomous a staff member also needs to acquire various capacities, dispositions, reflective skills and types of understanding. Professional autonomy is

2.3. Towards reflective skills in staff development

Within the framework of experiential learning it is possible to understand the changes in teachers and students. Learning is seen as a cyclic process of integrating concrete experiences, reflective observation, abstract conceptualisation and active experimentation into balanced and holistic understanding. The changes do not take place automatically; they require courage to encounter problems, solve them and evaluate personal action. They also imply cooperation with colleagues. (Niemi & Kohonen, 1995).

Developing the learner's awareness on all these aspects is suggested by experiential learning theory (Kolb, 1984) to which is added a mode of learning item from the viewpoint of metacognitive strategies (Kohonen, 1992).

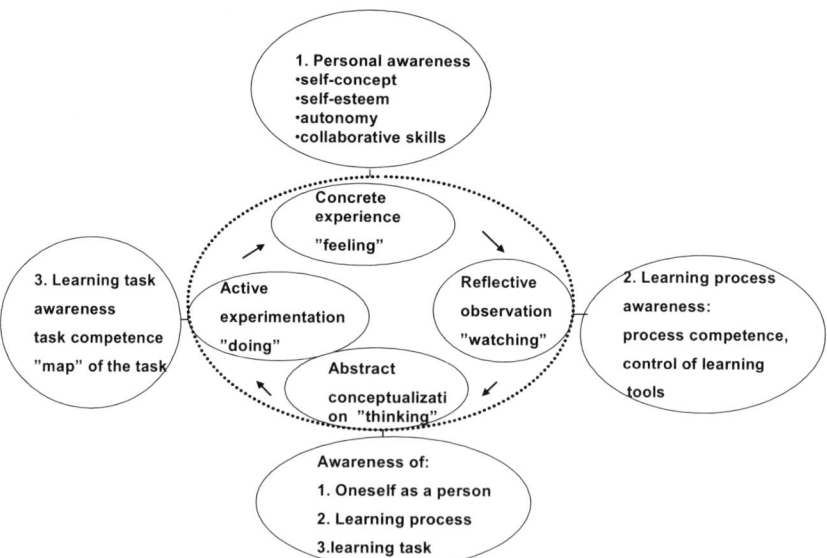

Figure 1. The theoretical frame of developing metacognitive strategies through experimental learning and reflection (Kolb, 1984; Kohonen, 1992).

Developing learning strategies, of which metacognitive strategies are of utmost importance to a reflective practitioner, can also be called 'learning to learn.'

2.4. Reflective skills as a key element in professional identity development

The purpose of reflective work is to integrate beliefs and images, program knowledge and classroom experiences both on personal and collective level. New understandings are created in the intra- and interactive processes by sensitising existing constructions, analysing them, soliciting conflicting perspectives and resolving the conflicts into new, better constructions. Thus reflection is very individual and personal on one hand; on the other hand it is collective by nature. The dialogue between the personal and collective should take place all the time. The awareness of this process as well as its intensity is flexible. For this process the teacher, as well as the student needs time and space. These factors are stressed also in the expertise development (Tynjälä, Nuutinen, Eteläpelto, Kirjonen & Remes, 1997).

The collective nature of teacher reflection can be compared to the efforts of physicists to explain the universe by the chaos theory. According to this and system theories as well, linear phenomena in nature are exceptions, non-linear are normal physical phenomena. Moreover, the most interesting things take place on the outskirts of the phenomena (Senge, 1990; Mäenpää ,1997). The teacher should become able to notice weak signals and either strengthen or weaken them. To become sensitive to weak signals the teacher needs networking with colleagues, community and the matters relevant to education. A parallel phenomenon to sensitising unseen but existing things is the arts: poetry, painting and music are examples of making implicit explicit.

3. Observations on reflective practices

This chapter describes an empirical research and its results. The target group was teacher trainees in vocational and university subject teacher education (No. = 54). The material was collected first in 1995-96 (40 trainees in the university program with the average age of 24), and second in 1997-98 (14 trainees in vocational education, the average age being 42). The descriptions of empirical practices in the following chapters concern mainly vocational teacher education.

University subject teacher education offers 35 credits in 9 months and is full time studies for students from different faculties. Vocational teacher education is directed to adult, graduated experts in different fields of sciences. A great number of them study part time. They are e.g. university and polytechnic in-service teachers, often doctors or licentiates, masters or engineers who work without the so-called teacher's licence. Gaining the pedagogical competence can last from 1 to 3 years. The slow track, the 3-year program consists of about 120-160 contact hours a year. Part of them can be replaced by e-learning at home. The so-called teaching practice, which is central in this study, is included in the contact hours above.

Compared to the pedagogical short courses of university teachers, the vocational programme is more comprehensive aiming at general teacher's competence, which is beyond the needs of a university teacher, whereas short courses are more concise. Reflective practices in both programmes can be equal.

3.1 Individual Reflection

The teacher trainees in the School of Vocational Teacher Education are encouraged to keep daily journals of their learning. They are asked not to write down what they were taught or told, but to write down key words that help them to return to the ideas and experiences, which they considered somehow important or puzzling. The emphasis is put on recognising their feelings in the course of actions. "What made me happy, what irritated me?" This is because recognising feelings and their causes is an important part of reflective practice. Similarly, incidents, developmental discussions, learning outcomes, feedback from different agents etc. are important material to their learning portfolios.

Later on they can read their notes and reflect on their experiences and find more theoretical views in reference books. They are asked to glance over the notes at the end of every week to make a fuller picture of the ideas occurring to them. They need their notes on several occasions:

- After contact hours they are often expected to write an assignment with help from their notes and reference literature. They are asked to include their own reflections in these essays. This is to connect the theories with their own experiences and practice.

- The first teaching practice unit is devoted to learner observation. Besides learning to know the students, the practice aims at learning how to collect empirical data about learners and how to interpret them. Thus after the "facts" comes the "fiction", i.e. they explain what they think went on in the classroom. The observations and interpretations are discussed in the seminar. After the sessions the trainees write short post-seminar reflective essays.

- The second practice unit emphasizes the learning contexts and observations of professionals at work. The trainees are asked to describe the study contexts and curricula, the learning conceptions and the values of the institutions as well as to write of their ideal teachers and learners. They are asked to analyse their own strengths and weaknesses as well as they can. This period is also finished with a seminar.

- The last practice unit emphasises building up and running a continuum of 16 to 32 hours of teaching. It is preceded by a written core plan giving theoretical reasons for their choices of teaching and evaluation methods as well as of materials, differentiation, motivation and organising the teaching process. Planning makes teaching well-prepared and systematic. To remind of the real life in classrooms, they are asked to notice the situational events in them as well. They are advised to ask some core questions themselves after every lesson (Fig 2).

The order of the units described above was changed in 2003, so today the trainees start with the observation of learning contexts, then go on to learner observation and finally have actual teaching practice.

After the last practice period the trainees write summative essays to conclude from the experiences written in learning journals. In the seminar each trainee also reflects orally on the incidents and phenomena, especially on those which touched them emotionally. These sessions are organised according to the model of collective reflection, sometimes called group dialogue.

The trainees are encouraged to discuss the themes described below with their supervisors in the schools and their co-trainees who audit the lessons. Also teacher trainers from the school of vocational teacher education audit a few lessons and give feedback.

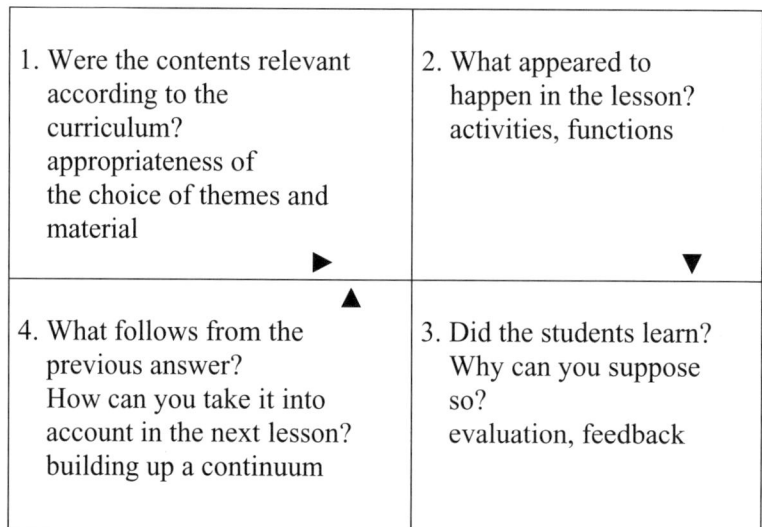

Figure 2. Reflective practice after lesson work: a matrix

This kind of systematic auditing and reflection gives ample material for portfolio work. Practice systems of this kind are possible to be built in actual teacher education programs, whereas short training courses should choose only the best practices.

3.2. Observations

The following analysis of the trainees´ essays (in the university language teacher program) reveals the contents and quality of their thinking. It is also interesting to see how they argue their views. The following categories of 1) self-understanding and professional identity, 2) pedagogical content knowledge, 3) educational purposes and values and 4) curricular goals, materials and evaluation, were relatively easily formed.

Self-understanding and professional identity. Teacher trainees seem to be much concerned with their images as teachers (34 %). They reflect on their past school days, their present qualities and skills, and imagine what they will be like in the future. Having unrealistic or vague expectations in the beginning they become more realistic and able to self-analysis, when time goes on.

I wonder how I will get on as a teacher. She must be able to do anything. (September)

A good teacher is expected a lot. It is impossible to meet all demands. I think I am good spirited, have a sense of humour, I am considerate, encouraging, honest, friendly and have a sound self-image. However I feel that I have a lot to learn before I am a good teacher. (the same person in February)

Pedagogical content knowledge is teachers' internalised procedural knowledge. 63 % of all trainee statements fell into this category. It was divided into sub-categories in the following way:

1. Subject matter knowledge can be widened to comprise metacognitive awareness of learning tasks and learning processes, including teaching methodologies of the subject in question, subject knowledge and its adaptation to learning purposes. The statements deal with this matter telling e.g. how a novice teacher comes to notice that

everything is not as easy in the class as it seems on paper (October)

I understood how important it is to see the globality of ideas in planning. Later I noticed that I had tried to invent too complicated methods.... However, there is no right or wrong way to teach. (March)

2. Knowledge of general pedagogy is an important area that students may be worried about. It includes classroom management processes and teaching strategies in general. The statements referred to the ability to motivate as an essential teaching skill as well as to managing the classroom control even in unpredictable situations.

3. Knowledge of learners and specific learning contexts ranges from working groups, communities and cultures to the school district management. This category includes the knowledge of learning environments and learners with different abilities and interests:

I try to think of my students as individuals... The most difficult thing is to notice the diversity of individuals in the class, as all do not learn in the same way. (March)

Educational purposes and values should be clear to all, and defined in every trainee's portfolio at least (Personal educational philosophy). Teaching is not tricks; it is meaningful action with a purpose. It ranges from educational values to the concepts of cognitive, emotional and social learning. Although these issues might seem distant to novice teachers, they reflect on the aims of education as well:

Our aim is to educate and support human growth. We should integrate this purpose and put it into practice in our work. (March)

Curricular goals, materials and evaluation. Since the National Board of Education in Finland has given freedom to schools and teachers to compose their curriculum's within the general national framework, it is important that trainees are aware of choices and alternatives:

All teaching, every lesson must be meaningful… It is traditional to follow a textbook but, fortunately, a teacher may use her own judgement and choose other material as well, if it is more suitable. (February)

I think it is fine that teachers can follow their own plans, set goals for themselves. On the other hand, you must decide everything yourself. But you can also work with colleagues. (March)

In tertiary education in Finland there is freedom of research and teaching. In universities and polytechnics there is, however, some kind of framework in every field of science concerning the contents and forms of teaching.

3.3. Dimensions of individual reflection

Although the self is strongly emphasized in the teacher trainees´ texts, there is also reflection about the backgrounds, interests and abilities of their students as well as their ways of seeing things. The trainees are either idealistic or doubtful at the beginning, and their conceptions of themselves as teachers are generally superficial and vague. The more they get experienced, the more concrete and realistic they become when thinking of their strengths and weaknesses.

How does systematic reflection influence the process of both pre-service and in-service teacher identity development in general? First, teachers become aware of themselves, not only as persons but as professionals. Second, along with the growing awareness, they become more daring to see themselves realistically. They do not try to hide their weaknesses, but they see them as the targets of conscious development and improvement. They also understand that they can emphasize their strengths. The learning process has thus become less painful and more realistic.

Third, reflecting on educational and psychological theories as well as the theories of their expertise domains produces more integrated knowledge. Further, changing meanings is an active process. It means active involvement in sharing and understanding. Changes appear to be a cyclic process of integrating concrete experiences, reflective observations, abstract conceptualization and active experimentation into a balanced and holistic understanding (Kolb, 1984). The changes to take place require courage to encounter problems, solve them and evaluate personal action.

Fourth, reflecting on learning and actions is a democratic process. Since the teacher trainees involved in the research have different backgrounds as teachers: some of them have a several-year teaching experience, others are true novices in the profession, and they could all start reflecting from the point where they were. The level of reflective skills varied from simple descriptions, to practical conclusions and even to deep analyses and understanding (Nissilä, 1999).

Changes imply cooperation with colleagues. Thus there is another important aspect, in addition to personal individual reflection. It is collective reflection and collective tacit knowledge.

3.4. Collective reflection

In collective **dialogue** there is free and creative exploration of complex and subtle issues, a deep listening to one another and suspending one's own views. By contrast, in **discussion** different views are presented and defended, and there is a search for the best view to support the decisions that must be made at this time. Dialogue and discussion are potentially complementary, but most teams lack the ability to distinguish between the two and move consciously between them. (Senge, 1990).

Team learning involves dialogue skills: learning how to deal creatively with the powerful forces opposing productive dialogue and discussion in working teams. The so-called defensive routines on one hand protect us from threat or embarrassment, but on the other hand, in doing so, prevent us from learning. (Argyris, 1992). For example, faced with a conflict, team members frequently either smooth over differences or speak out. Yet the very defensive routines that thwart learning also hold potential for fostering learning, if we can only learn how to unlock the energy they contain. It is the inquiry and reflection skills that enable the team to release this energy, which can then be focused in dialogue and discussion.

Team learning also needs practice. In fact, the process of learning is through continual movement between practice and performance. Despite its importance, team learning remains poorly understood. That is why it is most important to try to find out systematic features in team learning.

Bohm (Bohm & Peat, 1992) suggests that discussion is something like a Ping-pong -game. The purpose of a game is usually to win. A sustained emphasis on winning is not compatible with giving priority to coherence and truth.

On the other hand Bohm suggests that in dialogue a group accesses a larger pool of common meaning, which cannot be accessed individually. The whole organises the parts rather than trying to pull the parts into a whole.

3.5. How to enhance collective reflection

From the very beginning of their studies the teacher trainees in vocational teacher education are informed about the principles and philosophy of teacher education. During their contact teaching hours they are expected to share ideas and experiences. When they are "thinking aloud" of the phenomena they meet in theory or practice and when they share experiences and memories with each other, they activate their minds and start connecting theory and practice.

They are also informed about the differences of collective dialogue and discussion. Though involving individual skills and areas of understanding, team learning is a collective discipline. It also involves mastering the practices of dialogue and discussion, the two distinct ways the team's converse.

In the reflection seminars (described in 2.1.) the teacher trainees have their diaries and/or their summative essays with them. They first report of the events and their experiences. Second, they tell about their most positive and negative experiences and explain why they consider them either good or bad. This usually makes all the participants attentive and makes them compare their own experiences to the reports that they hear. Briefly, the stories activate their emotions. Participant's should suspend their assumptions and give the priority to each rapporteur, but after the report they should communicate their assumptions freely. The result is a free exploration that brings to the surface the full depth of people's experiences and thoughts, and can move beyond their individual views.

The teacher trainees are reminded of the importance of individual reflection. Collective reflection and dialogue are usually based on and intertwined with individual reflection. Consequently, the trainees are expected to write post-seminar reflections telling how their viewpoints were changed and how they evaluate the sessions (cf. Boud, Keogh & Walker, 1985).

The purpose of the collective dialogue is to go beyond any individual understanding. People are no longer primarily in opposition, nor can they be said to be interacting, they are rather participating in this pool of common meaning, which enables them to be involved in constant development and change.

3.6. Observations

In the following analysis collective reflection is approached through the observations of the team leaders (professors) and the post-seminar texts written by the teacher trainees.

There seems to be three basic conditions that are necessary for collective dialogue (Senge, 1990):

- All participants must suspend their assumptions, literally to hold them "as if suspended before us".
- All participants must regard one another as colleagues.
- There must be a facilitator who holds the context of a dialogue.

These conditions contribute to allowing the "free flow of meaning" to pass through a group, by diminishing resistance to the flow. Hot topics, subjects that would otherwise become sources of emotional discord and fractiousness become discussable.

Suspending the assumptions. Bohm argues that once an individual "digs his heels" and decides "this is the way it is", the flow of dialogue is blocked (Bohm & Peat, 1992). Suspending the assumptions is difficult because of the very nature of thought, which deludes us into a view that "this is the way it is". Collective discipline of holding assumptions suspended allows the team members to see their own assumptions more clearly. It also makes them daring to express their own conceptions:

We have noticed that the best gain in our studies has been our common reflection on learning and teaching from many points of view. .... Multi-subject teacher trainees enrich our views with their different aspects. ... People were very talkative and spoke out the things as they are.

Seeing each other as colleagues. Collegial attitude is critical to establish. In dialogue people actually feel as if they were building something, a new deeper understanding. Seeing each other as colleagues and friends is extremely important. We talk differently with friends. As the dialogue develops, team members will find this feeling of friendship developing even towards others with whom they do not have much in common. What is necessary is the willingness to consider each other as colleagues. It establishes the sense of safety in facing the risk of being sincere.

In telling about my experience in teaching practice I can be quite open. I have not met such things that I could not talk about. I do not either feel that I should be quiet for fear of criticism. ...I found it difficult to get started with my story, but then time flew when I was listening to interesting points of view. On the other hand, I think the emotional attitudes were talked about too much.

Colleagueship does not mean that you have to share the same views. On the contrary, the real power of seeing each other as colleagues comes into play when there are different views. But the payoff is great. Choosing to view "adversaries" as "colleagues with different views" has the greatest benefits.

Although we are very different in many respects, there are common features in us all. We all take teaching seriously and understand the responsibility connected with it. ... I was so open that I hope that the group understands that what we talkabout is confidential. I also could learn from other people's experiences.

According to (Bohm & Peat, 1992) hierarchy is antithetical to dialogue. Thus, everyone involved truly wants the benefits of dialogue more than he wants to hold onto his privileges of rank.

A facilitator who holds the context of dialogue. In the absence of a skilled facilitator, our habits of thought continually pull us towards discussion and away from dialogue. We believe in our views and want them to prevail. We are worried about suspending our assumptions publicly. We may even be uncertain if it is psychologically safe to suspend all assumptions - not to lose our identities (Senge, 1990).

The facilitator of a dialogue session carries out many of the basic duties for a good process facilitator. These functions include helping people maintain ownership of the process and outcomes. He also must keep the dialogue moving. As if he walks a careful line between being knowledgeable and helpful and yet not taking on the "expert" mantle that would shift attention away from the members of the team, and their own ideas and responsibility.

But the facilitator does even something more. He influences the flow of development simply through participating. For example, after someone has made an observation, the facilitator may remind of an opposite observation which also may be true. When the facilitator says what is needed and only what is needed at each point of time, it deepens other persons' appreciation of dialogue.

The group leader fosters open and warm atmosphere ... We were sitting in a circle in the room. Little by little we started to free ourselves from stress and excitement ... I also want to learn to become a group leader in this way.

As teams develop experience and skill in dialogue, the facilitator can gradually become one of the participants. Dialogue emerges from the "leaderless" group once the team members have developed their skill and understanding. In dialogue a kind of sensitivity develops which goes beyond what we normally recognise as thinking (Bohm & Peat, 1992).

Balancing collective reflection and discussion. In team learning discussion is the necessary counterpart of dialogue. In discussion decisions are made, in dialogue complex issues are explored. A learning team masters movement back and forth between dialogue and discussion. Failing to distinguish their different rules means that the teams usually have neither dialogue nor productive discussions (Senge, 1990).

If dialogue articulates a unique vision of team learning, reflection and inquiry skills may prove essential to realising that vision. Just as personal vision provides a foundation for building shared vision so do reflection and inquiry skills provide a foundation for collective reflection, i.e. dialogue and discussion. Dialogue that is grounded in individual reflection and inquiry skills is likely to be more reliable and less dependent on particulars of circumstance, such as the "chemistry", the personal dynamics among team members.

4. Recommendations

Reflection seems to be a vital tool in teacher and staff development. To make individual and collective reflection systematic it is advisable

- To keep a daily or weekly learning journal (diary, protocol), to read the notes at certain intervals and compose written or oral summaries. This helps to understand the theories dealt with, the practical occurrences in the real teaching situations and acquire pedagogical content knowledge as well.

- Autobiographies and recollections of past school days help the novice teachers to become aware of their preconceptions. This consciousness is necessary before creating new constructions.

- To internalise the core competences of teaching profession, all kinds of reflective and argumentative practice should be encouraged. The themes may range from self-image to the goals of education, from conceptions of learning to personal teaching philosophy and teaching as a profession, from violence in schools to the sociology of educating, not to forget the teaching methodologies and subject-related themes.

- To develop the kind of sensitivity that goes beyond what we normally recognise as thinking we have to concentrate on collective dialogue in team learning. This sensitivity is a fine net capable of gathering the subtle meanings in the flow of thinking as opposed to our individual nets gathering the coarsest elements of the stream.

- To make collective reflection, i.e. dialogue profitable and inventive, the facilitating skills of teachers and team leaders should be developed to the direction of democracy in collective reflection.

References

Argyris, C. (1992) *On organizational learning*. Blackwell, Cambridge, Mass.

Bohm, D. & Peat, D. (1992) *Tiede, järjestys ja luovuus* (Science, Order and Creativity), Gaudeamus, Helsinki.

Boud, D., Keogh, R. & Walker, D. (eds.) (1985) *Reflection: Turning Experience into Learning*, Kogan Place, London.

Calderhead, J. (1996) *Teachers, beliefs and knowledge* In: Berliner, D.C. & Calfee, R.C. (eds.), Handbook of Educational Psychology, Macmillan, New York.
Kagan, D. (1992) Professional growth among preservice and beginning teachers. *Review of Educational Research*, Vol. 62 No. 2, pp 129-169.

Kohonen, V. (1992) Restructuring school learning as learner education: toward a collegial school culture and cooperative learning, In: Ojanen (ed.) *Nordic teacher training congress: chanllenges for teacher's profession in the 21st century.* Research Reports 44, pp 36-59, University of Joensuu, Joensuu.

Kolb, D. (1984) *Experiential learning. Experience as the source of learning and development*, Prentice Hall, Englewood Cliffs, NJ.

Meijer, P., Beijaard, D. & Verloop, N. (2002) Examining teachers´ interactive cognitions using insight from research on teachers´ practical knowledge, In: Sugrue, C. & Day, C. (eds), *Developing Teachers and Teaching Practice*, pp 162-178, RoutledgeFalmer, London.

Mäenpää, J. (1997) *The core competence awareness of groups in learning business organizations*, University of Oulu, Oulu.

Niemi, H. & Kohonen, V. (1995) *Towards new professionalism and active learning in teacher development: empirical findings on teacher education and induction*, A 2/1995, University of Tampere, Tampere.

Nissilä, S-P. (1999) *From Individual towards collective reflection in preservice teacher education*, Paper presented in ISATT conference in Dublin, Ireland.

Richardson, V., Anders, P., Tidwell, D. & Lloyd, C. (1991) The relationship between teachers´beliefs and practices in reading comprehension instruction. *American Educational Research Journal*, Vol. 28 No. 3 pp 559-86.

Shulman, L. (1987) Knowledge and teaching: foundations of the new reform. *Harvard Educational Review* Vol. 57 No. 1, pp 1-22.

Senge, P.M. (1990) *The fifth discipline*, Doubleday/Currency, New York.

Schön, D.A. (1983) *The Reflective Practitioner: How Professionals Think in Action*, Basic Books, New York.

Tynjälä, P., Nuutinen, A., Eteläpelto, A., Kirjonen,J. & Remes, P. (1997) The Acquisition of Professional Expertise – a challenge for educational research, *Scandinavian Journal of Educational Research*, Vol. 41, Nos. 3-4.
Van Driel, J.H., Verloop, N. & De Vos, W. (1998) Develeoping science teachers´pedagogical content knowledge, *Journal of Research in Science Teaching* Vol. 35, No.6, pp 673-695.

White, J. (1990) *Education and the good life*, University of London, London.

Zeichner, K. & Gore, J. (1990) Teacher socialization. In Houston W. (ed.) *Handbook of Research on Teacher Education*, pp 329-348, Macmillan, New York.

Chapter 12

The Pedagogical Academy – a way to Encourage and Reward Scholary Teaching

Pernille H. Andersson and Torgny Roxå

1. Abstract/summary

The main objective of The Pedagogical Academy (Hammar Andersson et al., 2002) is to encourage pedagogical development at LTH. Another aim is to reward good, ambitious and quality-conscious teachers. This system of rewards will also initiate an improvement of teaching and learning among these teachers leading to a more professional approach in their teaching. The objective is also to get the teachers at LTH to contribute to the pedagogical debate and that the departments hopefully will provide even better teaching and take a greater responsibility for their employees' skills as teachers.

All university teachers, except postgraduate students, at LTH can apply to The Pedagogical Academy. All applicants participate in a workshop "How to write a teaching portfolio" during which they write their teaching portfolio (Apelgren & Giertz, 2001; Seldin, 1997). During the workshop, the teachers reflect upon teaching and learning and their own development as teachers. The main method used during the workshop is smaller groups having a dialog about these matters.

The teaching portfolio is assessed and scrutinized by colleagues, who are already in The Pedagogical Academy, as well as a pedagogical expert and a representative from the student union. The applicants are also interviewed by the assessment group. There are six criteria concerning scholarly teaching

(Boyer, 1990), which have to be fulfilled to a certain degree. If the teachers pass this assessment they receive the "Excellent Teaching Practice" (ETP) and an immediate rise in salary. The department will also receive an increase in their grant.

The Pedagogical Academy involves many parts of the organisation at different levels. The management level is interested in ways to encourage and stimulate a development in the organisation, which result in a higher level of knowledge of student learning and more pedagogical skilful teachers. The management level shows this intent by spending money as means of steering the pedagogical development. It also involves the teachers on a personal level. They develop as professionals and increase their knowledge of teaching and learning when they participate in the workshop and have their teaching portfolios assessed.

The Pedagogical Academy has been operating a year and a half and the experience, so far, shows that now there is more focus on questions of pedagogical issues. The dialog concerning a scholarly teaching and learning perspective is also increasing. Through the portfolios, where the teachers describe their teaching, the picture of teaching and learning at LTH becomes more obvious. This is helpful in the future planning of the pedagogical development at LTH.

The main problem is that at present there is no method to measure how the teachers interact with the students in real life. Other critical points are whether there is a sense of ownership of The Pedagogical Academy at LTH and the willingness of the teachers to go through this process.

2. Aims

Institutions providing higher education have to optimize two processes. Research, in order to acquire new knowledge of and competence in different subjects and disciplines, and the education of students. Many members of the academic staff chose their career because they have a great interest in a subject, and the academic career takes place in the area of research. As a consequence, most members of academic staff focus more on and put more effort into their research than their teaching.

The Pedagogical Academy was established in order to stress the equal importance of research and teaching at LTH. The system was developed within the Breakthrough project. The goal of this project is to implement a culture of scholarship of teaching and to contribute to a paradigm shift from a focus on teaching to a focus on learning in the teaching activities at LTH (Barr & Tagg, 1995). In this process the Academy has an important role.

The features of the Pedagogical Academy are described below.

- Good, ambitious, quality-conscious lecturers will be rewarded by a certificate of competence and an increase in salary. These lecturers and other ambitious teachers will be an indication that LTH is investing in good teaching and that there are goals to aim at.

- The departments from which lecturers have been admitted to the Academy will be deemed to have better capacity to provide good teaching. Moreover, if the department in question actively supports its lecturers in applying for and obtaining this certification, it is believed that, in the long run, they will find it easier to recruit and retain good teachers, and thus good students. For this reason, such departments will receive an additional financial contribution for every employee who achieves this certification. This system is based on what today is called *docentur*, i.e. achieving the grade of senior lecturer or reader.

- The aim of the system is to initiate positive development, in which it is clear that it pays to invest in good, carefully prepared teaching. This in turn will lead to professionalism in teaching, i.e. good teaching is documented and scrutinized, and thus acts as a springboard for further development.

- The certified lecturers are assumed to be able to contribute to pedagogical development at LTH. This may be realised through active participation in LTH's pedagogical debate and development, and by acting as mentors for other teachers.

With these aims, LTH is striving to develop the teaching culture in line with the present thoughts presented by many researchers and investigations concerning modern higher education (Boyer, 1990; Healey, 2000; Kreber, 2000; Abrahamsson, 2001; Fransson & Wahlén, 2001).

It is all about the quality of the students' learning and their abilities and knowledge when they graduate from the university as new engineers. Because the conditions for higher education have changed fundamentally in resent years, the way of looking upon teaching and learning and its aims must be discussed (Bowden & Marton, 1999).

Two main features in the Academy are important in this development. First, the teachers write a statement of their teaching and learning philosophy in their teaching portfolios, and give examples of how they apply their teaching philosophy in real teaching situations. This gives the teachers an opportunity to summarize their development and present thinking about student learning. This hopefully gives them more knowledge of their present standard in these issues and a more solid ground to develop their teaching skills from.

Second, the criteria the teachers shall meet to be admitted to the Pedagogical Academy and get the benefits of being a teacher with the Excellent Teaching Practice (see chapter 2) set standards for university teachers.

- that the applicant bases his/her work on a learning perspective,

- that the personal philosophy of the applicant constitutes an integrated whole, in which different aspects of teaching are described in such a way that the driving force of the applicant is apparent,

- that a clear development over time is apparent. The applicant should, preferably, consciously and systematically have striven to develop personally and in pedagogical activities,

- that the applicant has shared his or her experience with others, with the intention of vitalising the pedagogical debate,

- that the applicant has cooperated with other lecturers in an effort to develop his or her teaching skills, and

- that the applicant is looking to the future by discussing his or her future development, and the development of pedagogical activities.

3. Rewarded teachers

Teachers that are accepted in The Pedagogical Academy receive the Excellent Teaching Practice (ETP) and a monthly increase of their salary of 140 Euro. In order to encourage the departments to focus on the teachers' competence in teaching and learning, the department where a ETP teacher is employed gets an increase in its funding.

ETP teachers are expected to be mentors for other teachers who want to apply to Pedagogical Academy, and to be assessors in the application process.

At LTH, the most important roles of the ETP teachers are their effort to stimulate discussion of teaching and learning in the departments' every day life and to be persons known for having a certain interest and knowledge about teaching and learning issues.

4. Process being an ETP-teacher

To be able to apply to The Pedagogical Academy and receive the ETP, the teachers have to participate in a workshop on writing teaching portfolios. With the participation in the workshop, the teachers demonstrate their willingness to share practices and to communicate with colleagues about teaching and learning, which are important criterions for being an ETP teacher. After the workshop, the teaching portfolio is handed in for assessment, if the teacher thinks it is ready to pass the assessment. The teaching portfolio is read and scrutinised by a group of assessors and a scrutinizer. After an interview with the applicant, it is decided whether to approve or reject the application. If an application is rejected, the teacher can continue to work on the portfolio and apply again later. All applicants are offered feedback and advice from the pedagogical expert in the group of assessors.

5. Impacts in organisation

The main aim of The Pedagogical Academy is to stress the importance of competence in teaching and learning at LTH and the importance of related issues at LTH. Another aim is to stimulate development in this area both on the organisational level and at the individual level.

5.1. Organisational levels

Without evidence from systematic evaluations or results from research, it is noticeable that there is more focus on the question of teaching and learning at LTH now than before. Employees from many departments describe how they are more interested in discussing questions concerning teaching and learning than they were before.

The criteria stimulate the debate about teaching and learning, because they are discussed among the academic staff and must be understood by the teachers. Thereby, pedagogic issues are lifted to a higher level of awareness in the everyday discussions in the departments. This kind of discussions is one of the most important methods to develop the teaching and learning culture at LTH. It is creating a community of practice (Wenger, 1999).

The Pedagogical Academy also stimulates the development of different communities of practice among the teachers through the networks that is often created around the Pedagogical Academy. Nothing is done from a central level at LTH to activate the teachers which have received the ETP. It is up to the ETP teacher themselves to find ways of building networks and cooperate. At present, groups that try to form opinions on questions that have an impact on the teaching and learning activities at LTH are formed.

5.2 Personal levels

The teachers that go through the process of writing a teaching portfolio and have it assessed in this system also go through a development process. There are yet no systematic studies of this process and of what the personal learning outcome is, though there are plans of doing a study on this. During the workshop, the teachers write down their own reflections. During the writing of the portfolios. Often the teachers describe the process as very interesting because they have to think about and describe their own thoughts about teaching and learning. When they link this to the description of their practice as teachers they often get a clearer picture of their development, which often is very useful. Some teachers describe that they get more self-confidence as teachers and after having gone through this process they participate to a higher extent in the public debate about teaching and learning at LTH.

One assumption is that in the process of writing the teaching portfolio, the teachers will get their practical knowledge about teaching and learning turned in to professional knowledge, if they not yet have been through this process. By making the descriptions of their teaching activities and link them to pedagogical theories in the portfolio and use adequate theories and conceptions, their understanding of their own teaching will probably be more professional. It is also easier to share experiences with others if the teachers use the same expressions about phenomena in teaching and learning. The teachers gain new knowledge by reflecting upon their own practice. To have this reflective and scientific attitude to teaching and learning, leads to a culture of scholarship of teaching and learning (Boyer, 1990)

6. Implementation and credibility.

The implementation of the Pedagogical Academy has been an important issue to deal with.

The Pedagogical Academy has been developed by a group of five lecturers and professors from different departments at LTH and a pedagogical expert. This project group is responsible for the development of criteria, instructions and procedures for the assessment of pedagogical competence according to the basic ideas of The Pedagogical Academy. The representation from different departments in the group has been crucial for the implementation of the system in the organisation.

Another important factor in getting approval and support for The Pedagogical Academy at LTH was that the heads of the departments were constantly informed during the development process. The heads of the departments were also asked to select the first teachers who applied to The Pedagogical Academy.

The credibility of the system is still questioned. At present, the relevance of the criteria and the assessment processes are questioned. It is questioned if the rewarded teachers really are the ones that deserve it. This has also raised the question of what competence in teaching and learning really implies.

7. Evaluation and research.

Until 2003 no systematic and scientific studies on the impact of The Pedagogical Academy on the development of LTH have been carried out. During the autumn of 2003, two research projects started. One of them investigates the assessment process in the Pedagogical Academy to deal with the questions of competence in teaching and learning and what really is rewarded and assessed.

There is also an evaluation project that investigates how well the Breakthrough Project has achieved its aims and what impact this and The Pedagogical Academy has had on LTH.

The two projects will be reported during 2004 and afterwards the Pedagogical Academy at LTH will be further developed.

References

Abrahamsson, B. (2001) *Careers through promotion and recruiting*, (in Swedish), The National Agency for Higher Education, Stockholm.

Apelgren K., & Giertz, B. (2001) *Teaching Portfolios*, Uppsala University, Uppsala.

Barr, R. B. & Tagg J. (1995) *From Teaching to Learning – A New Paradigm for Undergraduate Education*, Change (Nov/Dec) pp 13-25.

Bowden, J. & Marton F. (1999) *The University of Learning*, Kogan Page, London.

Boyer, E. L. (1990) *Scholarship Reconsidered. Priorities of the Professoriate*, The Carnegie Foundation, New Jersey.

Fransson, A. & Wahlén, S. (2001) *New conditions for learning in higher education*, SOU 2001:1, Stockholm.

Hammar Andersson P., Olsson T., Almqvist M.. et al. (2002) *The Pedagogical Academy – a way to encourage and reward scholarly teaching*, Proceedings 10:th conference Improving Student Learning, Oxford.

Healey, M. (2000) Developing the Scholarship of Teaching in Higher Education: a discipline-based approach, *Higher Education Research & Development*, Vol. 19 No 2.

Kreber, C. (2000) How University Teaching Award Winners Conceptualise Academic Work: some further thought on the meaning of scholarship, *Teaching in Higher Education*, Vol. 5 No 1.

Seldin P. (1997) *The Teaching Portfolio – A Practical Guide to Improved Performance and Promotion/Tenure Decisions*, Anker Publishing Company, Inc., Boston.

Wenger, E. (1999) *Communities of Practise, Learning, Meaning and Identity*, Cambridge University Press, Cambridge.

Chapter 13

Portfolio – Why, what and how

Søren Hansen, Säde-Pirkko Nissilä, Claus Spliid and Anders Ahlberg

1. Abstract

Teaching portfolios are increasingly common in Nordic academic environments. In this article we discuss how and why portfolios are compiled, and describe current examples of academic portfolios. The limitations of teaching portfolios are outlined.

2. Introduction

Although teaching portfolio writing has been used by individuals and organisations for a long time, it now seems that this tool will become considerably more widespread in Nordic university environments. This coincides with the turnover in many European countries from voluntary to compulsory pedagogical training of university teachers, and with urgent needs of faculties to evaluate pedagogic merits in a fair and systematic way (documentation of professional achievements). But, portfolios can provide so much more. The writing of portfolios ideally leads to self-reflection, and provides the starting point for personal and professional development and improved classroom process awareness. In other words, ideally it makes teachers drop mechanistic or intuitive performances and adopt conscious choices of classroom action/interaction in tune with student abilities. These advantages can be achieved by students, as a combined learning and examination activity (learning portfolio).

The start of portfolio-writing can be an obstacle, but once the initial threshold is passed, there is commonly a "creative flow", fuelled by the growing insight which has indeed been development. Although portfolio-writers include those who have kept a good record of their teaching achievements and reflections, more commonly teachers try to backtrack decades of development, through memories. Quite commonly this reveals punctual evolution of the teachers' performance, driven by frustration when teaching appears inefficient (Kugel, 1993). In these instances portfolio-writing offers a stimulating possibility for teachers to analyse the rationale behind their own changes. However, the human memory is not an objective record of events (witness observation experiments clearly demonstrates this). It may be further distorted (rationalised) by teachers' experiences and standpoints throughout their careers, and by the way their teaching has been appreciated among colleagues and students. There may be difficulties to keep up self-discipline when portfolio-writing turns from the long-term retrospective stage to documentation of current teaching. On the other hand, from this stage and onwards one may design the documentation around the teaching activities so that they indeed verify the quality of the learning outcome. Further, as time goes by and the quality of course documentation is good, the "ill-documented past" will gradually fade in importance.

The poor documentation of teaching merits, compared to those of research, is a real problem for academics and faculties, which can be helped with the use of teaching portfolios (Apelgren & Giertz, 2001; Giertz, 2003). However, if one wants to invoke current ideas of academic scholarship, i.e., undividable intertwined research and teaching, then a teaching portfolio should be an integrated "extractable" part of each academics "portfolio of competence", rather than a stand-alone separate document.

3. Ways to compile a teaching portfolio

A portfolio is the collection of documents from which the author can compile different versions of portfolios for different purposes. Thus a portfolio is flexible. It is dynamic in the sense that it is being renewed all the time. Old documents can be removed and new materials fitted in instead. The focus should be on the newest documents. This building process is meant to continue lifelong.

The comprehensiveness can also vary according to the aim for which the portfolio is designed. A person can have 1) a teaching portfolio, which is usually comprehensive, very personal and private. Out of this a teacher can build 2) a showcase portfolio for recruitment or 3) a portfolio for an employer's use. It can, for instance, be accessible even on the www. In addition, a person can have 4) a learning portfolio which is meant to document the development of its author during a certain period of time.

A teaching portfolio is the collection of documents that are being updated all the time. The material collected can be sorted out so that the most recent and/or most important documents get the amplest space. This type of portfolio can consist of, besides the curriculum vitae and the documents of exams and the career, excerpts from the learning diary and teaching plans. Outcomes of learning (of one's own and one's students) and other documents should be included. Feedback from the students should be analysed and summarized. "Raw material" is usually too comprehensive to be included as such. Even the future expectations and visions have their place here. This kind of documentation offers the author an opportunity to reflect on different aspects of his life and work. It also causes developmental discussions with the head of the institution.

A showcase portfolio is produced for a certain purpose. The documents and other material are chosen to illustrate the aspects that are vital for that purpose. The lay-out is designed by the author. A personal touch is inspiring. If the author wants to produce a portfolio in the internet for the needs of the employer or as a part of his/her home page, the documents and data included in it should be negotiated. The author decides which data are included in this *public portfolio* and which in the teaching portfolio only.

As an instrument of evaluation the portfolio both guides and reports: it reveals the development in the long run, which is not easy to notice otherwise. It helps the teacher to become aware of his strengths and weaknesses. It also helps to see emotionally coloured incidents from a new point of view, in the perspective of time. This is the case e.g. in evaluating the student feedback and processing the teachers' self-assessment connected with it.

A prerequisite in compiling the portfolio is that all data presented must be attestable by different documents. Teachers could include the following kinds of data in their "tool bags":

- *Curriculum vitae* (more or less comprehensive) can be written freely. It acts as a kind of widened official CV, which also is added to the portfolio to document the text written freely.

- *Teaching philosophy*; a teacher is supposed to verbalize personal views on how students learn and how a teacher should support it. He should also give examples of the ways he has tried to put the philosophy into practice. These should be shown in the documents introduced later in the portfolio.

- *Acting and development as a professional*; this could include descriptions and documents of teaching: courses, lectures etc. As attachments the author could add specimens of plans, and self-assessment of the outcomes, as well as students' learning products and feedback. The products of learning can be texts, pictures and/or video- or audiocassettes. The teaching material that the teacher has produced can be illustrated by examples. Also the tests devised by the teacher and respective test results could be included. The student feedback is included analysed and summarised. If a teacher has collected feedback from his colleagues, this should also be in the portfolio (Kohonen, 1992).

A teacher can also document his further and continued education data, and those of all the activities that illustrate his development as a professional. Even books, seminars etc that have roused his interest and reflection could be mentioned here.

To the development as a professional, activities in different organisations also belong, especially in the educational field. Activities in organisations and international networking can be titled respectively.

Personal achievements are also collected into the portfolio. For instance the lists of publications and specimens of other writing activities as well as appearing on the public scene in different contexts can be included with or without reflection.

Hobbies and interests can also be presented in the portfolio. The author can create titles to his portfolio divisions according to the kind of things he wants to introduce on the pages. He can use colours, pictures, photos, poems, mottos etc. to make his portfolio as individual as a fingerprint.

4. Various roles of teaching portfolios

4.1. A portfolio as an instrument of development and recruitment

One of the foremost factors for success in an organisation is the staff working in it. Their professional growth promotes the development of the whole organisation. "While traditional organisations require management systems that control people's behaviour, learning organisations invest in improving the quality of thinking, the capacity for reflection and team learning, and in the ability to develop shared visions and shared understandings"(Senge, 1999).

Similar points of view about learning communities are expressed in the (OECD, 1992) which says that the new competencies of teachers include, besides the relevant subject matter skills, flexibility in thinking, self-directed, continuous learning, creativity, good communication, team work, risk-management and taking the initiative.

It necessitates a new orientation to teaching both to understand the teaching profession and the changes in it. Essential in it is the shift from teaching as a transmission of knowledge to transactional and transformational teaching whereby learners are facilitated to autonomous, active thinking and learning for themselves. This means in practice that when documenting interaction with students, the teacher will be able to see the pattern of interaction in his teaching contexts. When encouraging his students to write learning portfolios and diaries, he will learn something of their ways to conceptualise the subject matter. The expectation is that reflection will lead to growing autonomy of both the teacher and the students. The responsibility of learning is the student's; the teacher is the facilitator and the organiser of the learning opportunities.

Those changes cannot be achieved from the outside, but they are self-directed, growing from the inside. The constructivist learning conception, which regards the learner (either a teacher or a student) as a constructor of his own learning and as a team learner, emphasizes the mastery of metacognitive knowledge and skills: self-cognition as well as planning, implementing and evaluating one's own learning and behaviour. A portfolio-work and writing learning diaries is a good way to self-cognition.

In recruitment as well as in developmental discussions the portfolio is an important asset when considering a teacher's strengths and the qualities to be developed. Future expectations can also be seen in a natural perspective.

4.2. Portfolio-writing: reflection as a tool in course development

The ways in which teachers conceptualise their professional tasks depend on their basic philosophical orientation. Their prior beliefs and experiences are often stable, inflexible and difficult to change (Nissilä, 1997). If the teachers are expected to create some new practices in teaching or adopt a new learning conception, for instance, they should first become aware of what their present views are. It is important to make teachers´ silent knowledge, their implicit conceptions explicit, and facilitate the ability to learn from experience and theory through systematic reflection.

Reflective skills are needed in all learning and teaching. Besides increasing personal awareness, they are used as a tool in evaluating the theoretical and practical contents of the profession. The more deeply the concepts are understood, the deeper and more focused is the reflection.

The aim of reflection is to make the "unobserved observable". This process, which generates new understanding, is intrapersonal, interpersonal and interactive. It is very individual on one hand, and collective on the other (Senge, 1999). The dialogue between the individual and collective, i.e. inside a person and with colleagues and learning communities should be continuous. The intensity and flexibility of the dialogue can vary. The process needs time and space. For this process documents are necessary, and they could be composed into the form of a portfolio.

4.3. Portfolio as a learning diary

On the pages of his *learning diary* the learner writes down his doings, thoughts, feelings and the situations he has met. The diary is free in form and individual, either an intimate document of one's own thinking and emotions or a written personal document which can be shared partly or in full by another person(s). The first process in writing a diary is arranging the experiences and thoughts in the way that they can be verbalized. Verbalization helps the writer to learn something of the process that is going on inside him. It acts as the tool of increasing awareness. Thus learning diary often precedes or is intertwined with portfolio work.

A learning diary can also focus on some special area. A *lecture diary* can replace a written exam after lectures. It is expected to go deeper into the themes of the lectures, widening them with the help of literature and growing into a profound essay. It is usually marked in the same way as exams whereas an individual learning diary is not marked in the traditional way. Its value is greatest to its owner.

In conclusion, a teaching portfolio can be used as an instrument of the merit system and recruitment as well as an asset in professional and personal development. It is the self-assessment and reflection that support the growth of professional identity and self-respect.

5. An example of portfolio work from Aalborg University

5.1. Introduction

At Aalborg University staff development includes a pedagogical course which is compulsory to all Assistant Professors. The pedagogical principle of the course is that learning takes place in action and by keeping a reflective dialog up with supervisors, students and colleagues. Relevant theory is learned from workshops and self-studies. A portfolio is used as a tool for self-reflection and as documentation of teaching experience, focussed on the reflections on action and progression in teaching practice. The course material is mainly divided into two categories. The first contains inputs, ideas, theories and methods for teaching and supervising. The second contains facilitative material for self-reflection. Lists of reflective questions support facilitation, which has the purpose of making the participants, reflect on their own practice. In this way they continuously improve their teaching methods and curriculum. Facilitating by questioning central aspects of a teacher's practice is a very efficient way of implementing a learner-centred approach to staff development. The two supervisors of each course participant ask and discuss reflective questions.

5.2. What competences should and could be documented?

The main competence is to be able to reflect on ones own learning. The course participant must demonstrate that he/she is able to develop his/her own teaching methods through reflection. One way of doing that is to write down the objectives for the teaching before giving a lecture. After the lecture you reflect to which degree you have reached the objectives. This may be based on data collected in connection with lecturing. As a result of the reflection you formulate the objectives for the next lecture. From the portfolio it should then be possible to observe a progression both in terms of a change in teaching methods and in terms of reflections. A progression in reflections means that the portfolio writer demonstrates that he/she is continuously developing a competence in reflection of his/her own learning. This means asking the right questions, collecting the right data, making proper analyses and coming up with suggestions for future practice.

5.3. Who will read the portfolio? For what purpose are portfolio compiled?

The portfolio has tree main target groups. The self-directed work of the portfolio-writing teacher includes documentation of conscious, systematic ver-

166 · Portfolio – Why, what and how

balisation of the pedagogical progression. This also means using the portfolio as a tool for developing the consciousness, the personal pedagogical terminology and for mapping one's explorations. Another target group is the supervisors, and the third is the appointment committee which assesses the assistant professor when he/she applies for a position as an associate professor. The supervisors look for progression, and use the portfolio as a starting point for facilitating further reflection. The appointment committee looks for proof of teaching competences. The portfolio maybe separated into a working portfolio which is shown to the supervisors and a showcase portfolio which is shown to the appointment committee, because the target groups not are looking for different aspects.

5.4. How should portfolios be evaluated?

At Aalborg University the portfolio is mainly evaluated as a working portfolio, from which the Assistant Professors' progress and competences in reflection should appear. At the same time, it is also a showcase portfolio, documenting experience with teaching. Based on regular observations, evaluations and reflections documented by the Assistant Professor in the teaching portfolio, the supervisors compile a report at the end of the course concerning the Assistant Professor's development as a teacher.

6. What a portfolio does not provide

In the previous section it was described how the portfolio is used as a didactic mean to develop and evaluate the teaching of assistant professors. In the present section it will be questioned whether it is appropriate to use a written portfolio for self-reflection and documentation of competences.

The portfolio is a way to deal with reflection, but what kind of reflection? To understand the limits of the portfolio I want to introduce two different kinds of reflections. Schön differs between reflection – in – action and reflection – on – action (Schön, 1983). Reflection – on – action is something you do after the action has stopped. You look back on the action and analyse it. Is there something I should change in my future action? Portfolio writing works well in this respect. The portfolio monitors your own development. Reflection – in – action is something completely different. It is something you do throughout the action. It is a way of experimenting with action directly without having a clear objective. The portfolio cannot help you with that.

There are many ways to deal with reflection of ones own practice. It is something that develops as a personal method that, in time, should become an integrated part of ones learning style. In the beginning, a structured tool such as a portfolio might be helpful, but later it may seem too structured and rigid and many practitioners stop using it (Hansen, 2000). One reason for that is that there is a big difference between reflective thinking and writing. Although the writing process can be a way to engage in reflecting and thereby conceptualising your doing, it is also time-consuming and functions merely as a double take on what you already have been doing. An example may illustrate this. You know that sometimes the students lose their concentration in the middle of a lecture. You have some ideas about how to avoid this, but even though you have structured your lecture in a way to avoid it, it still happens from time to time. In the middle of the action written reflections is not very helpful, you need a good idea right now, so you start reflecting – in – action, which means that you start experimenting and reflecting on the spot. After the lecture you then reflect – on – action to find out what was going on and how it worked, but that will always be a double take on what happened – or what you *think* happened. This may lead to the next paragraph.

To be able to reflect on – action, you look back on your previous action. This implies that you know what actually happened. Schön has shown that often there is a difference between our espoused theories (what we think happened) and our theories – in – use (what actually happened). Humans are very good at filling out holes in memory with imaginations. The portfolio writer may thus "lie" to himself when he is trying to find the objectives to follow. However, this may be avoided by having a reflective dialog with peers or a supervisor.

Another limitation in the written portfolio is that it limits the writer's reflection to what can be expressed in words. Teaching and learning processes are much more complex than that. There might be a danger that the portfolio prevents us from using knowledge gained from other learning styles such as reflection – in – action, on the spot experimenting, intuition and improvisation. The problem in this respect is that the portfolio can only contain cognitive knowledge, which in some respects constitutes a very limited representation of reality.

Furthermore, reflection stops action, especially written reflection. It is a critical act more than a developmental act. This may be problematic in connection with creative problem solving where the objective is to get and develop ideas without instantly killing them with critical reflective questions.

For the reasons outlined above, I would like to stress the importance of having the portfolio as a merely written event supplemented with a reflective *dialogue* and *demonstration* of competent practice. In the example from Aalborg this means that both the Assistant Professors and the supervisors should spend more time *doing* than writing and reading. By observing the teaching of Assistant Professors, and by keeping a reflective dialogue, Assistant Professors and supervisors can improve their understanding of learning and teaching much better than by portfolio writing only.

References

Apelgren, K. & Giertz, B. (2001) *Och plötsligt var jag meriterad!* Rapport 27, Enheten för utveckling och utvärdering, Uppsala Universitet.

Giertz, B. (2003) *Att bedöma skicklighet - går det?* Rapport 2, Enheten för utveckling av pedagogik och interaktivt lärande, Uppsala Universitet.

Hansen, S. (2000) *"Vejledning og evaluering af den refleksive praktiker i det problemorienterede projektarbejde på ingeniørstudiet ved Aalborg Universitet"*, Aalborg Universitet.

Kohonen, V. (1992) Restructuring school learning as learner education: toward a collegial school culture and cooperative learning, In: Ojanen, S. (ed), *Nordic teacher training congress: challenges for teacher´s profession in the 21st century*. Research reports of the University of Joensuu 44, pp 36-59. University of Joensuu, Joensuu.

Kugel, P. (1993) *How professors develop as teachers*, Studies in Higher Education 18, pp 315-328.

Nissilä, S-P. (1997) Raising Cultural Awareness Among Foreign Language Teacher Trainees, *In: Byram, M. & Sarate, G. (Eds.), The Sociocultural and Intercultural Dimension of Language Learning and Teaching*, Council of Europe, Strassbourg.

OECD, (1992) *Schools and business: a new partnership. Paris: Center for Educational Research and Innovation.*

Senge, P.M. (1999) *The Fifth Discipline,* Random House Business Books, London.

Schön, D. (1983) *The reflective practitioner. How professionals think in action*, BasisBooks, Harper Collins Publishers.

Chapter 14

Continuous Assessment

Bertil Larsson and Anders Ahlberg

1. Abstract

Learner-centred teaching inevitably involves continuously ongoing context-specific assessment of students understanding, attitudes, problem solving abilities, etc. At LTH Department of Electro science students recognise this as an informal integrated part of the behaviour of appreciated teachers, who in turn claim they developed this rewarding strategy intuitively. The department has made classroom assessment methods familiar to its teachers, and formally integrated continuous student-feedback in all courses. Questionnaires show that in large classes (>80 students) positive effects of classroom assessment techniques are obvious to most students. In small classes (<30 students), the outcome is unclear, either due to well working subtle informal classroom communication (not obvious to students), or merely due absence of formative assessment.

To be efficient, formative assessment should be a "private" productive dialogue between teacher and class. It is therefore imperative from the schools´ perspective to monitor that formative assessment is ongoing without "eardropping" on the classroom dialogue. Such systems are possible to design, for instance by means of individual class-specific web-sites with restricted access and limited life-time. To secure continuity and robustness of the quality of each taught course, the final summative post-course evaluation includes the students and teachers evaluation of the formative assessment as a main element.

2. Aims

As a part of LTHs ambition to improve teaching and learning (project "Genombrottet"), one of its departments, Department of Electro science, decided to launch the pilot project "Operative assessment" aiming towards improved learner-centred education. The main idea behind this was that teachers' true understanding of their student's perspectives on course curriculum and activities was significant for "good teaching". The question was weather it was possible and useful to systematically integrate such monitoring of student views into the teaching system of a school (Roxå, 2002). In this context, it is imperative to distinguish formative assessments that continuously improve ongoing classroom activity (operative assessment) from conventional post-course evaluations, which report problems and outcomes to those outside the classroom after the course is finished (the school board, the student organizations, the university administration, the sponsors, etc).

3. Classroom assessment

Many teachers conduct classroom assessment intuitively and without thinking much about it. They may toss out questions in or before class which monitor deep understanding of the course context, they may read the facial expressions and body language of the students, or simply sample the moods of the students during coffee breaks. The problem is that not all university teachers do this. For instance they may have problems understanding informal off-class language, or sense they have too big a class to monitor. Or, they may simply not appreciate the virtues of learning-centred teaching. Further, teachers spontaneous intuitive monitoring of their students abilities may not always address urgent and specific questions related to the course curriculum. There are, however, well-established assessment methods available. A wide spectrum of efficient classroom assessment techniques (CATs') were established and has been accepted globally (Roxå, 2002; Angelo & Cross, 1993; Black & Wiliam, 1996 & 1998; Davis, 1993 and www.siue.edu). Although they mostly are used to monitor course-related knowledge and skills, they may also be specifically aimed at other course aspects, for instance to assess critical and creative thinking, attitudes and values, or learner reactions to instructions and group activities. The use of classroom assessment techniques are normally not particularly time consuming for teachers or students, and do give teachers an opportunity to clarify, repeat, or change perspective of central concepts during subsequent class. Each CAT provides a positive loop, ideally leading to deeper student understanding of course topics prior to the introduction of next new concept (Fig. 1).

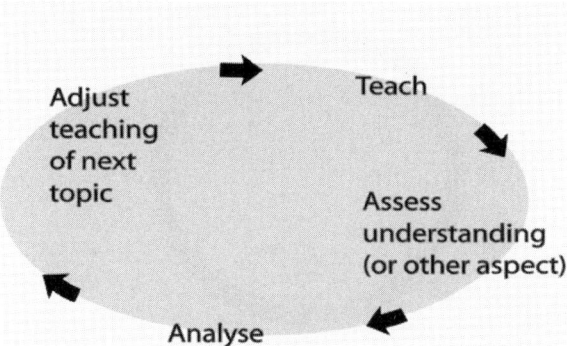

Figure 1 Loops of classroom assessment need to be context specific, frequent and easily administered to impose positive impact on teaching and learning.

In this way, teaching of each concept takes its starting point from the students´ pre-existing conceptual framework. To be an efficient tool in improving student learning, classroom assessment inevitably must be continuously ongoing, student-centred, teacher directed, and mutually beneficial for students and their teacher. It further needs to be context-specific, i.e., adapted to the learning situation of the specific part of the curriculum taught at the moment (Figure 1).

4. Technical aspects

Classroom assessment can be conducted in simple ways within the classroom for instance by students anonymously leaving answers in the teacher's mailbox before leaving the room. Assessment may also be conducted around the clock on class web sites with restricted access (teacher and class only). However there is a difficulty in the systematisation of classroom assessment, as it is not aimed at those outside the classroom door. This can readily be solved by routinely asking students and teachers in the post-course summative assessment whether operative assessment has been going on, and to what degree it has had impact on teaching (Figure 2).

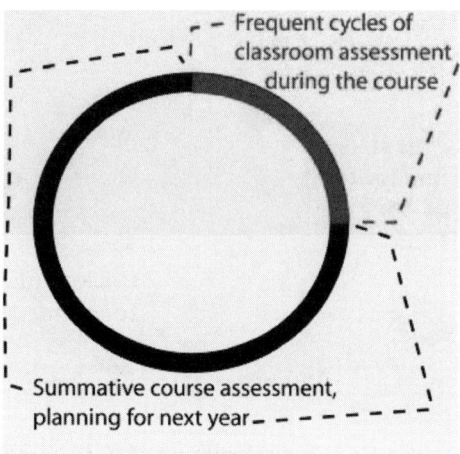

Figure 2 Relation between summative and formative assessment.

5. The pilot study

Our pilot study includes operative assessment of all courses taught at the Department of Electro science during the fall of 2002, including a full range of freshmen to advanced students. Electro science is a large academic unit which recently was formed by merging of three research units, the Department of Applied Electronics, Department of Telecommunication Theory and Department of Electromagnetic Theory. Typically classes are large (up to 150 students), and 20 teachers are involved in teaching. The introduction of systematic classroom assessment ("Operative assessment") coincides with an administrative reorganisation of the teaching staff, which was necessary to secure and promote high quality courses. The courses are now supervised by a group rather than by a single teacher to ensure long-term follow-up and course improvement.

The teachers at Electro science have recently been introduced to the principles of classroom assessment techniques, and were during the first semester (spring 2002) urged to find and modify assessment techniques which suite their classroom situations best. During this trial, a pedagogical consultant (AA) was available to meet questions or hesitance regarding the introduction of classroom assessment. After a 6-month period of trial, CAT activities have been evaluated.

6. Results

Increased classroom assessment lead to typical advantages, i.e., increased student motivation as students realise that their teachers do care about their learning, and optimal "student knowledge growth" as teachers keep better track of the knowledge level and quality in their classes during the courses. In this context, increased student motivation has stimulated students to contribute feedback truly useful to their teachers.

The main question we ask ourselves in the longer perspective is whether continuous syn-course assessment (operative assessment) can be systematically imposed on a teaching organisation, or, if such pedagogic virtues are inevitably linked to the personal development of individual teachers regardless of administrative setting.

The results are based on interviews with teachers and questionnaires to the students after finishing the courses.

The introduction of the concept to the teachers was over all well received. There was consensus regarding the need for this kind of assessment although there was some hesitation regarding the implementation. At a second meeting this topic was discussed and new ideas were added to the ones already in use. After a couple of weeks we made some interviews on the impact so far. Some had succeeded in monitoring the learning outcome and hence reorganised the course plan. Other had noticed a lack of understanding but chosen not to respond to this, and we also saw one or two who ignored the whole concept on the presumption that it is each students' responsibility to learn the course, and as regards the teacher this kind of assessment only delays and disturbs the general plan. The most positive signal from the teachers' interviews was the consensus that "This has a value of it's own for the teacher". It did indeed enhance the quality of the courses where it was performed at full extent, which is a guarantee for the continuance outside the frame of a pilot study.

The students were asked two questions after finishing the course:

Operative assessment means that the teacher is continuously monitoring student learning of each concept in order to adjust the course plan when needed. This may for instance include assignments, discussions, anonymous notes etc.

1) Have you experienced that operative assessment has occurred?
2) If so, what is your comment on the outcome of the operative assessment?

174 · Continuous Assessment

Student responses to question 1) are shown in Table 1 and Fig. 3. A positive response is more obvious for large classes, typically more than 80 students, than for small classes, typically less than 30 students.

Course	Number of students	Did experience Operative assessment	Did not experience Operative assessment	No answer
1	27	9	14	4
2	126	85	16	20
3	80	12	3	65[1]
4	41	23	1	17
5	20	12	0	8
6	22	6	7	9
7	120	41	5	74

Table 1. Outcome of students view on operative assessment.

1. Few students were reached by the questionnaire on this course.

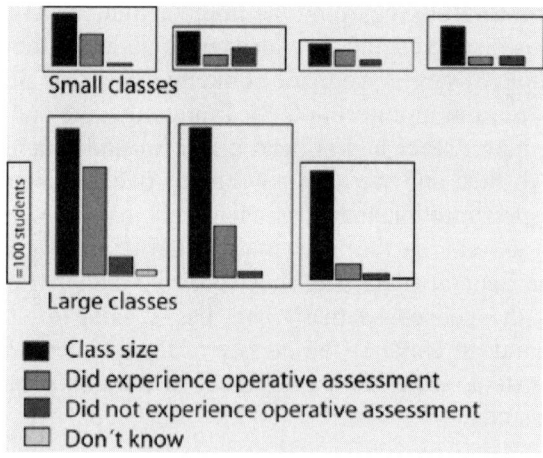

Fig. 3. Graph based on Table 1.

The explanation to this is the fact that in small classes the communication between the students and the teacher is much easier and direct. Students in small classes do not always know when operative assessment is done because the teacher actually know each student's capacity and arrange the teaching accordingly. For large classes the benefit of operative assessment is more obvious for the students. They see the interest shown from the teacher and the outcome of the assessment that they are asked to reflect on. It is a better strategy to let the students provide the way to improve the result. Their own reflection on the learning process is very rewarding in the long run.

7. Conclusion

This study shows that operative assessment as described will enhance learning, especially in large classes. When encouraged the teachers find it personally rewarding and valuable for the class. We believe that a framework at departmental level effectively can support ongoing operative assessment regardless of course type and teacher personality. Obvious pitfalls include teacher's fear of diverting from the original course plan, or not realising that sticking to the plan may hamper learning. We are aware of the constraint of a fixed time schedule. Maybe the schedule should focus more on the topics to cover, and less on chapters and weeks, so that each topic has to be penetrated and abandoned only after sufficient learning has been achieved.

8. Acknowledgment

We thank teachers and students involved in this pilot study for their input. Our friends involved in LTH-Genombrottet are acknowledged for their support and feedback throughout this study.

References

Torgny R, (2002) *Personal communication.*

Angelo, T.A. & Cross, K. P. (1993) *Classroom assessment techniques. A handbook for university teachers* (2nd Ed.), Jossey-Bass Publishers, San Fransisco, pp 427.

Black, P. & Wiliam, D. (1996) Meanings and Consequences: A Basis for Distinguishing Formative and Summative Functions of Assessment. *British Educational Research Journal*, Vol. 22, pp. 537-48.

Black, P. & Wiliam, D. (1998) *Inside the Black Box: Raising Standards Through Classroom Assessment*. Phi Delta Kappan Online Journal, Retrieved: August 23, 2004, from [http://www.pdkintl.org/kappan-/kbla9810.htm.

Davis, B.G. (1993). *Tools for Teaching*. San Francisco: Jossey-Bass.

List of contributors

Anette Kolmos, Aalborg University, Denmark

Ole Vinther, Copenhagen University College of Engineering, Denmark

Pernille Andersson, Lund Institute of Technology, Lund University, Sweden

Lauri Malmi, Helsinki University of Technology, Finland

Margrete Fuglem, Norwegian University of Science and Technology, Norway

Erik de Graaff, Technical University of Delft, The Netherlands

Gunilla Jönson, Lund Institute of Technology, Lund University, Sweden

Torgny Roxå, UCLU, Lund University, Sweden

Hans Peter Christensen, Technical University of Denmark, Denmark

Vidar Gynnild, Norwegian University of Science and Technology, Norway

Johanna Naukkarinen, Helsinki University of Technology, Finland

Säde-Pirkko Nissilä, The School of Vocational Teacher Education, Oulu Polytechnic, Finland

Søren Hansen, Aalborg University, Denmark

Claus Spliid, Aalborg University, Denmark

Palle Qvist, Aalborg University, Denmark

Anders Ahlberg, Lund Institute of Technology, Lund University, Sweden

Bertil Larsson, Lund Institute of Technology, Lund University, Sweden